A Botanist's Vocabulary

A
Botanist's
Vocabulary

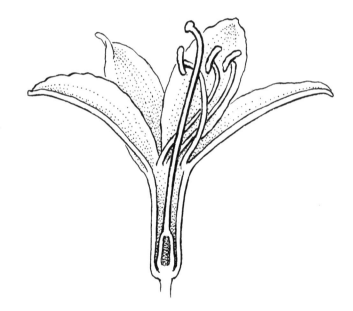

1300 TERMS EXPLAINED
AND ILLUSTRATED

SUSAN K. PELL and **BOBBI ANGELL**

Timber Press • Portland, Oregon

Published in 2016 by Timber Press, Inc.

The Haseltine Building
133 S.W. Second Avenue, Suite 450
Portland, Oregon 97204-3527
timberpress.com

Printed in China
Fourth printing 2020

Cover and text design by Anna Eshelman

Library of Congress Cataloging-in-Publication Data

Pell, Susan K., author.
  A botanist's vocabulary: 1300 terms explained and illustrated / Susan K. Pell
and Bobbi Angell.—First edition.
    pages cm
  ISBN 978-1-60469-563-2
  1. Botany—Dictionaries. I. Angell, Bobbi, 1955– author. II. Title.
  QK9.P44 2016
  580.3—dc23   2015029699

A catalog record for this book is also available from the British Library.

Contents

Introduction

6

The Glossary

10

Recommended Reading

224

Introduction

THERE IS AN INHERENT CURIOSITY among gardeners and naturalists that manifests itself in our conversations and actions. We are the ones making note of a plant label in a botanical garden to go back home to learn more or possibly locate one for our own garden. To us, a fallen blossom on a path begs to be pulled apart and examined up close with a hand lens. A rare plant catalog or a new book about an interesting genus are likely to captivate us for hours. Gardeners and naturalists discuss their observations, noting the first buds that emerge in the spring, describing an unusual feature on a particular plant, sharing cultivation tips for difficult habitats, or giving directions to obscure but botanically wonderful locations. We apply common names as well as Latin names to discuss the plants we encounter or cultivate along the way. We describe the colors, shapes, and textures of the plants, the growth forms and fruit characteristics, but often inadequately, perhaps not knowing the proper word to describe a particular feature, or knowing one word to describe it when our companion or manual uses a different word. We may recognize that there is no such thing as a typical flower, no basic leaf shape or growth habit, but we often cannot come up with the descriptive term for the complex and the unusual. Such terms help us to categorize and organize the world in which we are so intimately involved. Learning and applying the correct term leads to a far better appreciation for the incredible diversity of plants, enables us to communicate our knowledge, and allows us to access an even more technical and in-depth body of literature to satisfy our interests in the botanical world.

We have attempted to define terms used by botanists, naturalists, and gardeners alike to describe plants. We have simplified and clarified as much as possible to encourage the use of a common language. The included terms mostly refer to plant structures and come from the horticultural and botanical literature and practice. Many, perhaps most, terms are not easily defined or illustrated. If they were, the botanical kingdom would not be as rich and engaging as it is. With infinite variety, petals and sepals sometimes adhere to each other to attract pollinators or facilitate pollination; male and female reproductive parts may fuse to form intricate unified columns; fruits have peculiar, sometimes complicated, mechanisms of seed dispersal. There are terms that apply only to a particular group of plants, such as orchids, grasses, or irises. Some apply to whole plants or ecosystems, while others are visible only under a microscope. Please wander through the book to recognize the easily applied terms and learn a few unusual ones, but also use the book as a reference when you are stumped by a field guide or a strange-looking fruit. We hope your newfound knowledge helps you gain an even greater appreciation for the world of plants.

Bobbi Angell
Susan K. Pell
May 2015

The Glossary

a-
prefix meaning without, lacking; e.g.,
apetalous

abaxial
lower surface

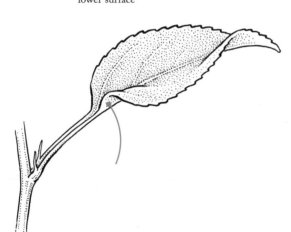

abcission
separation of one structure from another,
as with leaf from stem or petal from flower
receptacle; the result of cells breaking down
at the base of the falling structure

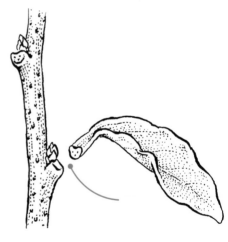

acaulescent
lacking an above-ground stem
ANTONYM caulescent

accessory fruit

fruit whose flesh is partially or wholly derived from non-ovary tissue (often from the receptacle); e.g., strawberries (*Fragaria*)

achene

small, dry, indehiscent fruit derived from a unicarpellate pistil; e.g., clematis (*Clematis*)

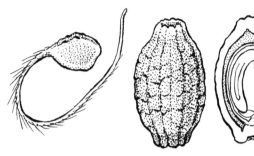

accrescent

growing larger after flower maturity, most often applied to the calyx

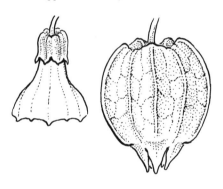

achlorophyllous

lacking chlorophyll; e.g., parasitic plants such as Indian pipe (*Monotropa uniflora*)

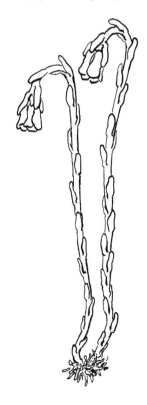

acerose

three-dimensionally needle-shaped

SYNONYM acicular

acicular

three-dimensionally needle-shaped

SYNONYM acerose

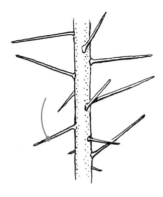

acidophilous

acid-loving, as with many plants that grow better in acidic soil

acorn

the nut fruit of oaks (*Quercus*), which is embedded in a scaly cup and has a single seed

acropetal

growing toward the tip of the shoot or root

ANTONYM basipetal

actinomorphic

having multiple planes of symmetry such that any line drawn through the middle produces two mirror-image halves, usually applied to flowers

SYNONYM radially symmetrical, regular

ANTONYM bilaterally symmetrical, irregular, zygomorphic

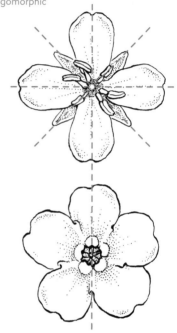

aculeate

having prickles

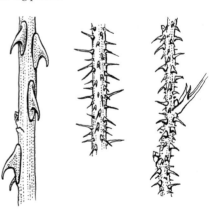

acuminate

tapering to a point with concave margins

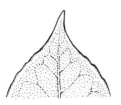

acute

pointed with the sides straight or nearly so and forming a ≤90° angle, applied to both the base and apex of leaves

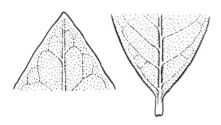

adaxial

upper surface

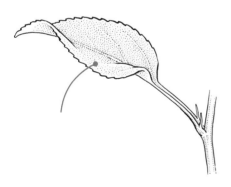

adherent

unlike structures stuck together but not fused

adnate

unlike structures fused together, as stamens to petals

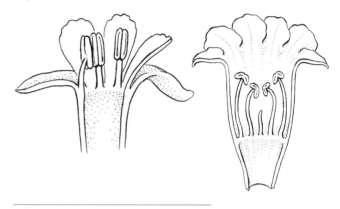

adventitious

structures arising in areas other than those in which they normally occur, most often applied to roots arising from shoots or leaves

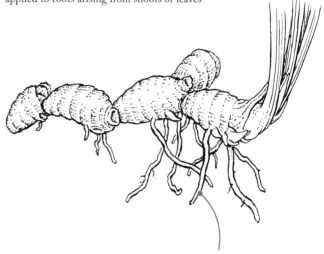

adventive

relatively recently escaped and spreading non-native species; less established than naturalized species

aerial

occurring or produced above the soil or water; e.g., aerial roots

aerial bulb

bud usually produced in leaf axils that can give rise to a new plant; e.g., plantlets on cycad trunks

SYNONYM bulbel, bulbil

aerial root

root produced above ground, such as those produced along the climbing stem of poison ivy (*Toxicodendron radicans*)

aestivation, estivation

arrangement of perianth parts in bud; see also vernation

afterripening

period of rest (also called dormancy) that some seeds must go through before being able to germinate

agamospermy

production of viable seeds without fertilization, without sexual reproduction

aggregate fruit

formed from the fusion of multiple, separate unicarpellate pistils in a single flower, may consist of tiny versions of one of many different fruit types including samaras, drupes, achenes, follicles, etc.; e.g., raspberries and blackberries (*Rubus*)

SYNONYM etaerio

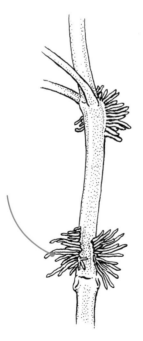

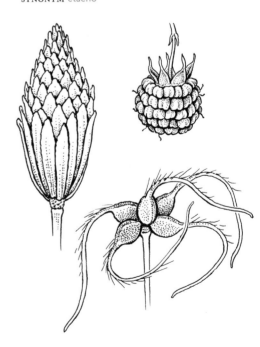

air plant

a plant that grows attached to another plant but which does not parasitize that plant, often specifically refers to epiphytic bromeliads in the genus *Tillandsia*

alpine plants

plants that grow in high-elevation areas above where trees grow (i.e., above the tree line); these are the plants traditionally and most commonly grown in rock gardens

alate

winged with expanded and flattened tissue; e.g., stems and fruits of winged elm (*Ulmus alata*) and stems of winged euonymus (*Euonymus alatus*)

alternate

1. occurring one per node, as with leaves interspersed on a stem; 2. occurring one following the other in a series, as with petals alternating with sepals in a flower when viewed from above or below

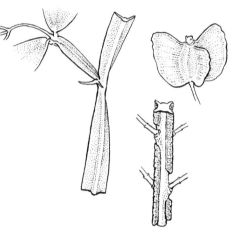

allelopathy

phenomenon in which a plant secretes compounds that interfere with the growth, reproduction, or continued survival of other plants around it; e.g., black walnut (*Juglans nigra*)

allopatric

occurring in different areas, as with two species whose distributions do not overlap

ANTONYM sympatric

alternate bearing

producing abundant fruit every other year, and minimal to no fruit in the alternating year

SYNONYM biennial bearing

alternipetalous

said of floral parts whose positions in the flower alternate with those of the petals

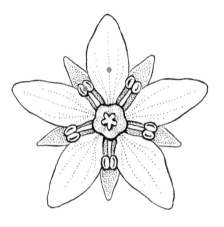

alternisepalous

said of floral parts whose positions in the flower alternate with those of the sepals

alveolate

honeycomb-like with neatly arranged depressions and ridges

SYNONYM faveolate, favose

ament

usually pendulous spike-like inflorescence of unisexual, sessile to subsessile, apetalous flowers

SYNONYM catkin

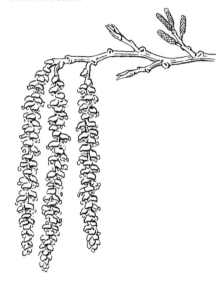

amplexicaul

clasping the stem without fully encircling it, as with a leaf, stipule, or bract

ampulla
bottle-shaped or spherical swelling

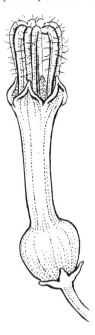

anastomosing
reconnection of branches resulting in the formation of a network; e.g., reticulate venation

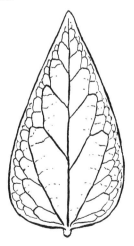

anchor root
adventitious root emerging from the lower part of a trunk and acting as structural support for a tree

SYNONYM brace root, prop root, stilt root

ancipital
flattened, with two edges

androecium
male reproductive portion of the flower, consists of stamens

androgynophore
stalk elevating the androecium and gynoecium above the perianth, as in passionflowers (*Passiflora*)

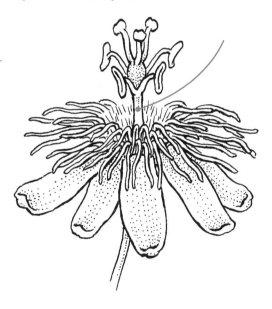

androphore
stalk bearing the stamens

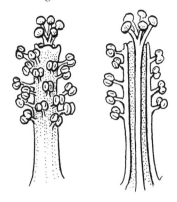

anemophilous
pollinated by wind

angiosperm
plant bearing flowers that have ovules inside of ovaries that develop into seeds inside of fruits

anisomerous
having an unequal number of parts in the floral whorls

anisophyllous
opposite leaves of unequal size and/or shape

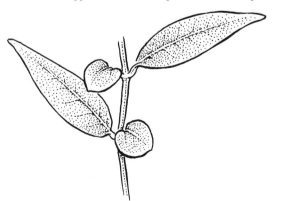

annual
plant whose entire life cycle occurs in one year: it grows from seed, flowers, produces seeds, and dies

annular
shaped like a ring, as in a nectary disk

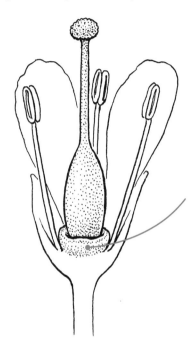

annulus
line of cells along the sporangium wall that plays an important role in sporangium dehiscence and spore dispersal in ferns

antepetalous, antipetalous

directly in front of the petals, as with stamens occurring opposite the petals, not alternating with them

antesepalous, antisepalous

directly in front of the sepals, as with stamens occurring opposite the sepals, not alternating with them

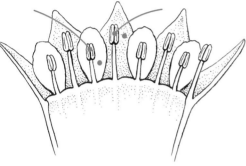

anther

pollen-bearing part of stamen

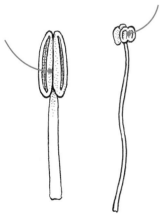

antheridium

(plural antheridia) male, sperm-bearing reproductive structure of ferns, lycophytes, and non-vascular plants

anther sac

one of typically two chambers inside of each anther, usually bears pollen

SYNONYM theca

anthesis

peak flower maturity, flower is open and sexually fertile

anthocarp

seed-bearing structure resembling and often mistaken for a fruit but for which the majority of the tissue is non-ovary (may be from such structures as a hypanthium or receptacle); e.g., rose hips (*Rosa*)

SYNONYM false fruit, pseudocarp

anthocyanin

blue, red, and purple pigments in plants, type of flavonoid, water-soluble

anthophore

stalk between the point of attachment of the calyx and the rest of the flower (corolla, androecium, gynoecium)

anthoxanthin

white and yellow pigments in plants, type of flavonoid, water-soluble

ant-plant

a plant that has a mutualistic relationship with ants

SYNONYM myrmecophyte

antrorse

pointing up or toward the apex

ANTONYM retrorse

aperture

opening, specifically refers to those in the exine of pollen grains

apetalous

lacking petals

apex

(plural apices) tip; opposite the base and furthest from point of attachment

ANTONYM base

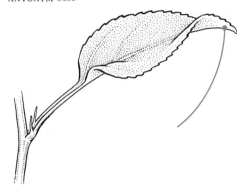

aphyllous

lacking leaves

apical

at or of the apex, as in apical placentation

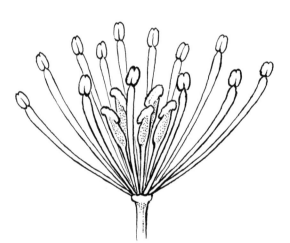

apical dominance

terminal bud's control over vertical growth of the main stem and elongating growth of branches by growing more vigorously than, and limiting, the growth of lateral buds; this phenomenon gives rise to the typical form of trees

apiculate

coming to an abrupt, short point

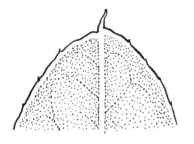

apiculum, apiculus

an abrupt, short point

apo-

prefix meaning apart, detached, or separate

apocarpous

gynoecium consisting of two to many separate, unfused carpels (simple pistils)

ANTONYM syncarpous

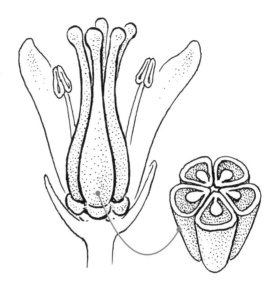

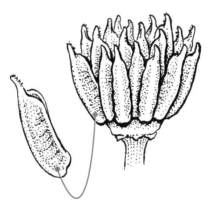

apomictic

reproducing asexually from the flower or fruit, often used synonymously with agamospermy

apomixis

asexual reproduction involving the flower or fruit, often used synonymously with agamospermy

appendage

structure arising from another, larger, structure

appressed

close against another structure but not fused to it

aquatic

plant growing in water seasonally or continuously

arborescent

tree-like in form but not a true tree; e.g., banana (*Musa*), which is an herbaceous plant with a trunk made up of leaf bases

archegonium

(plural archegonia) female, egg-bearing reproductive structure of ferns, lycophytes, and non-vascular plants

arctic

growing north of the arctic circle

arcuate

arching or curved; e.g., arcuate venation

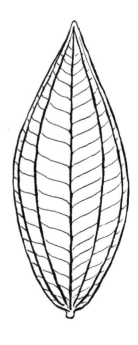

areola

small area distinct from nearby areas; in leaves, the space created by the joining of veins; in the cactus family (Cactaceae), the area on the stem that produces spines, glochids, flowers, etc.

aril, arillus

fleshy outgrowth of the funicle or hilum, subtending to encompassing a seed; e.g., lychee (*Litchi chinensis*) and yews (*Taxus*)

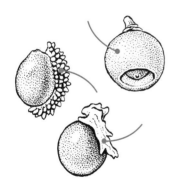

arista

a bristle, usually terminating a leaf or other structure

SYNONYM awn

aristate

terminating in a long bristle

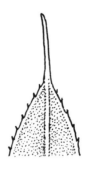

armature

prickles, spines, thorns, etc.

articulation

notch or joint, often the location at which two sides or organs separate

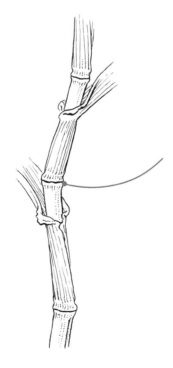

ascending

pointing or growing upward in an arching or curving fashion

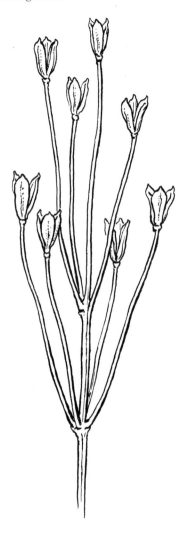

ascidiate

pitcher-shaped, as in the water-holding leaves of pitcherplants (*Sarracenia*)

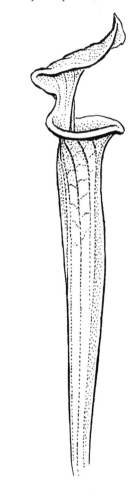

asepalous

lacking sepals

asexual

without sex, as with individuals or reproduction

asymmetrical

having two halves or sides that are unequal in size and/or shape, usually applied to leaf bases

SYNONYM oblique

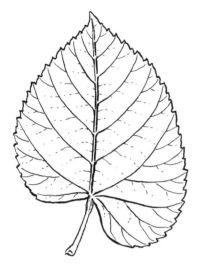

atropurpurea

dark purple

attenuate

gradually narrowing to a point

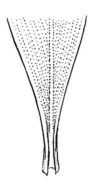

auricle

earlobe-shaped appendage, as in some leaf bases

auriculate

earlobe-shaped; having auricles

SYNONYM eared

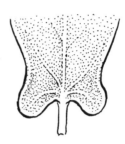

autogamy

self-pollination, self-compatibility

awn

a bristle, usually terminating a leaf or other structure

SYNONYM arista

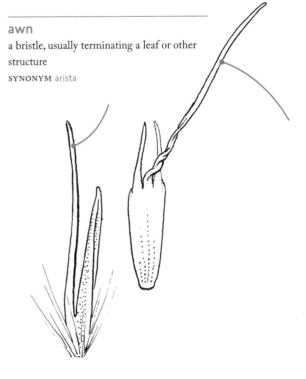

A

axil

upper angle of the junction between a stem and another stem, leaf, or reproductive structure

axile

1. of the axis; 2. attached to the axis, as in axile placentation

axis

central vertical portion of a structure, to which parts of that structure are often attached, as with an inflorescence and its branches or a flower and its whorls

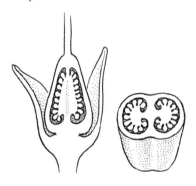

axillary

at the junction between a stem and another stem or other organ, most often used in reference to buds

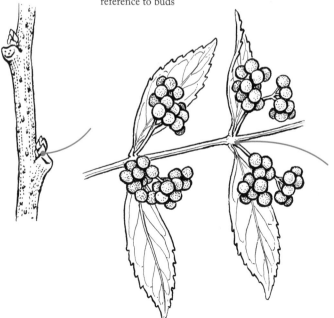

bacca

fleshy, indehiscent fruit with one to many
seeds embedded in a pulpy mesocarp,
endocarp not easily discernible; e.g.,
blueberries (*Vaccinium*)

SYNONYM berry

baccate

fruit that looks like a berry but may or may
not be a true berry; often applied to tropical
and unusual berries; e.g., avocado (*Persea
americana*)

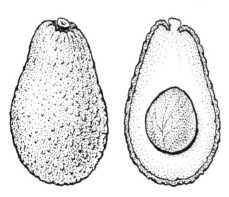

back bulb

old orchid pseudobulb that no longer has
leaves, often used for propagation

balausta

fleshy, indehiscent, many-seeded fruit
with leathery exocarp, derived from a
multicarpellate pistil; e.g., pomegranate
(*Punica granatum*)

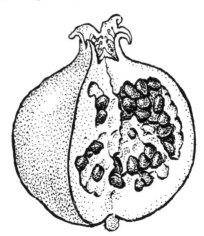

banner

flower petal typical of papilionoid legumes
in the bean family (Fabaceae), usually the
upper and largest petal; e.g., sweet peas
(*Lathyrus*), lupines (*Lupinus*)

SYNONYM standard, vexillum

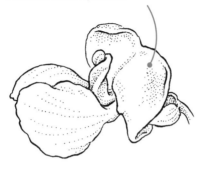

bark

the outer layer of woody stems, consists
of living phloem, cork cambium, and cork
(all tissues to the outside of the vascular
cambium)

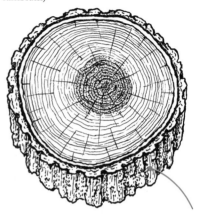

barbed

with stiff, sharp, retrorse (or less commonly
introrse) points

basal

at, attached to, or of the base, as with leaves
attached to the base of a plant

bare root

roots that would normally be surrounded by
soil are exposed, plants are often shipped in
this state to prevent the spread of soil-borne
insects and pathogens

basal placentation

ovule(s) attached at the base of the ovary in a simple pistil

basal plate

small stem portion of a bulb from which the roots grow downward and the leaves, inflorescence, and bracts grow upward

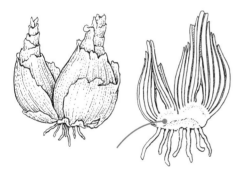

basal shoot

shoot growing from the base or roots of a tree or shrub, usually applied to those emerging from below ground

base

portion opposite the apex; closest to or at the point of attachment

ANTONYM apex

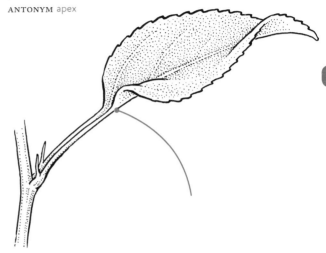

basifixed

attached at the base, as with filaments attached to the base of anthers; see also

dorsifixed, medifixed, versatile

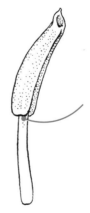

basipetal

growing toward the base of the shoot or root

ANTONYM acropetal

beak
an elongate tip or projection

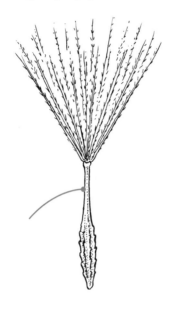

bean
the seed of a legume; also sometimes used to refer to the whole fruit, in which case it is being used as a synonym of legume

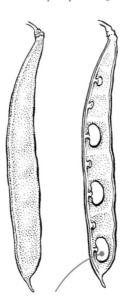

beard
clump of hairs or fringe of tissue that appears fuzzy, as on the central portion of the three outer tepals of some irises (*Iris*)

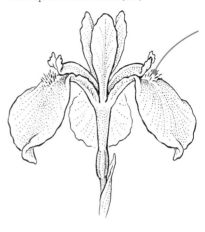

bearing
having or producing; most commonly applied to plants that produce edible fruits

berry
fleshy, indehiscent fruit with one to many seeds embedded in a pulpy mesocarp, endocarp not easily discernible; often applied incorrectly to any small fleshy fruit; e.g., blueberry (*Vaccinium*)
SYNONYM bacca

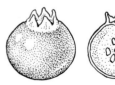

bi-
prefix meaning two

bicarpellate
having two carpels

bicolored

having two colors, usually applied to flowers

biconvex

shaped like a lentil, i.e., round and convex on both sides

SYNONYM lenticular

bicrenate

two-tiered scalloped margin where larger scallops have smaller scallops on them

SYNONYM doubly crenate

bidentate

having two teeth

biennial

plant whose entire life cycle occurs in two years: it grows from seed and produces leaves the first year, often as a basal rosette; it flowers, produces seeds, and dies the second year

biennial bearing

producing abundant fruit every other year, and minimal to no fruit in the alternating year

SYNONYM alternate bearing

B

bifid

split in two, as in some leaf blades

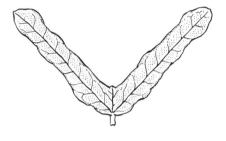

bifoliate, bifoliolate

having two leaves or leaflets

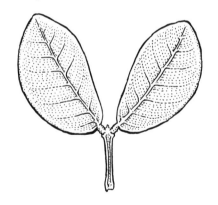

bifurcate

split into two branches

bilabiate

having two lips, as with some flowers; e.g., mint family (Lamiaceae)

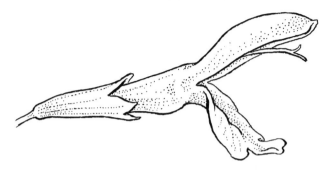

bilaterally symmetrical

having a single plane of symmetry such that only one line drawn through the middle produces two mirror-image halves

SYNONYM irregular, zygomorphic

ANTONYM actinomorphic, radially symmetrical, regular

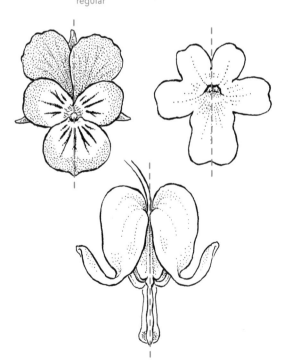

bilobed

having two lobes

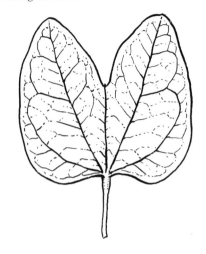

binomial

the two-part scientific name for a species, consisting of a genus (e.g., *Acer*) and a specific epithet (e.g., *rubrum*)

bipinnate

leaf that is pinnately dissected twice, leaflets arising along rachillas that themselves arise along a rachis

bisected

divided into two parts

biseriate

in two series

biserrate

margin teeth having teeth of their own, all pointing up toward the apex

SYNONYM doubly serrate

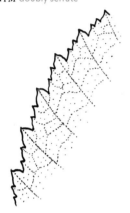

bisexual

having both female (egg) and male (sperm) reproductive cells in the same individual or reproductive structure

bitoned

having two tones of the same color, usually applied to flowers

black knot

large, dark, warty masses on the branches of *Prunus* spp. (especially cherry and plum) caused by the fungus *Dibotryon morbosum*

bladder

sac-like structure, often filled with air or liquid

blade

the usually broad and flattened part of a leaf or petal

SYNONYM lamina

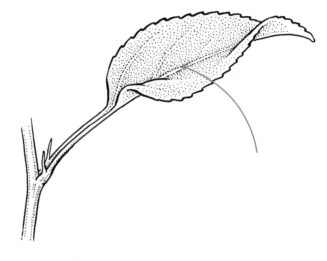

blind shoot

stem of a flowering plant that does not produce flowers, most commonly applied to roses (*Rosa*)

bloom

1. a flower or inflorescence; 2. gray-white waxy or powdery surface coating; 3. rapid localized growth of algae

blossom

a flower or inflorescence

bole

main stem or axis of a tree, between the roots and where branches begin to form the crown
SYNONYM trunk

bolt

grow rapidly, usually happens with seedlings or saplings and often occurs after a resource becomes newly available

bonsai

1. a woody plant, most commonly a tree, that is substantially reduced in size, usually accomplished through intentional manipulation under cultivation, but may also occur naturally under challenging growing conditions; 2. the Japanese art of growing dwarf woody plants

boot

remnant leaf base that remains attached to the trunk of some palm trees after the leaf dies

bough

a branch of a tree, usually applied to the larger branches

bourgeon, burgeon

a sprout or bud

brace root

adventitious root emerging from the lower part of a trunk and acting as structural support for a tree

SYNONYM anchor root, prop root, stilt root

brachyblast

stem with highly compressed internodes that usually bears the leaves and reproductive structure; e.g., ginkgos (*Ginkgo*), apples (*Malus*)

SYNONYM short shoot, spur

ANTONYM long shoot

bract

leaf-like structure subtending a flower or inflorescence or occurring in the inflorescence

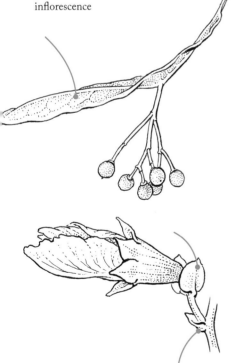

bracteate

having bracts

bracteose

having numerous or showy bracts; e.g., the inflorescence of dogwood (*Cornus florida*)

bramble

a prickly plant, usually specifically applied to raspberries and blackberries (*Rubus*), less commonly to their other relatives in the rose family (Rosaceae)

branch

1. stem arising from another stem; 2. the divarication of one structure into smaller versions or segments of that structure; 3. the action of producing a divarication, as in branches of trees or veins in leaves

branchlet

small branch

breaking

1. opening, as with flower or leaf buds in the spring; 2. ending, as with dormancy of a germinating seed

breastwood

new branches growing on an espaliered tree that usually must be pruned to keep the shape of the espalier

bristle

narrow, stiff hair

bud

immature flower, leaf, or stem still with its protective covering (bud scales, bracts, sepals, etc.)

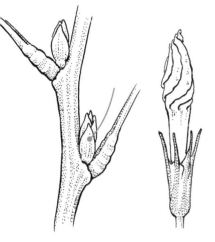

bulb

underground storage structure composed of buds attached to a small basal plate of stem tissue surrounded by fleshy leaf bases and bracts, majority of structure is leaf tissue; e.g., onions (*Allium*)

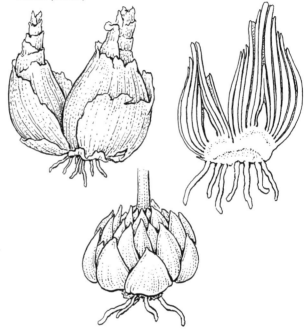

bud scales

small leaf-like structures on the outside of the bud, protecting the developing flower, leaf, or stem

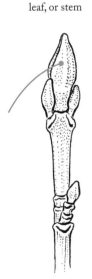

bulbil, bulbel

bud usually produced in leaf axils that can give rise to a new plant; e.g., plantlets on cycad trunks

SYNONYM aerial bulb

bulblet

small bulb, often growing from the base of a larger bulb

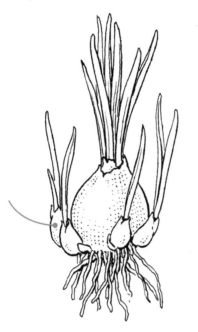

bullate

having a surface covered in smooth, rounded bumps

bundle scar

mark within the leaf scar on the stem from where leaf's vascular tissue was attached to the stem

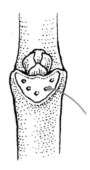

bur, burr

seed dispersal unit (may be seed, fruit, carpel, etc.) that is armed for sticking to fur for dispersal

burgeon, bourgeon

to send forth new growth rapidly, as with a shoot bud

burl

woody knot in the trunk, branches, or
roots of a tree, caused by something that
has damaged the tree (infection, disease, or
injury), prized for use in woodworking

bush

woody plant with multiple main stems,
usually shorter than a tree

SYNONYM shrub

buttress

widened base of tree trunk, most commonly
associated with trees in wet areas; e.g., bald
cypress (*Taxodium distichum*)

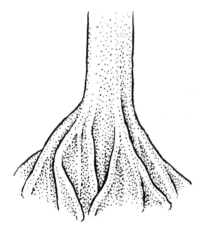

C

caducous

quickly deciduous, often applied to flower petals

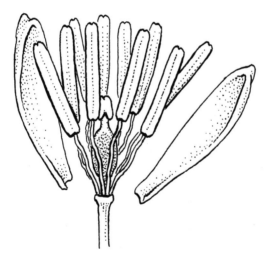

caespitose, cespitose

growing in dense clusters

SYNONYM clumped

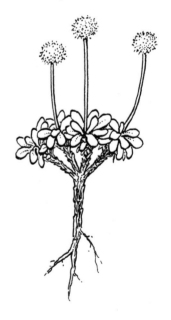

calcar

pointed, hollow appendage on a flower, often containing nectar and a projection from or a modification of the perianth

SYNONYM spur

calcarate

having a spur

SYNONYM spurred

calcareous

1. soils that are high in lime (calcium carbonate); 2. vegetation that grows on lime-rich soil

calceolate

shaped like a slipper, as in the pouch-like labellum of slipper orchids (Orchidaceae subfamily Cypripedioideae)

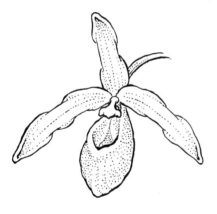

calcicole

plant that grows best in calcareous soil

SYNONYM calciphile, calciphyte

calcifuge

plant that does not grow well (or in some cases at all) in calcareous soil

calciphile, calciphyte

plant that grows best in calcareous soil

SYNONYM calcicole

caliper

diameter of a tree trunk at six inches above the ground, measured at 12 inches above ground when trunk diameter is greater than four inches; see also DBH

calloused

having a callus

callus

(plural calluses, calli) 1. hard, thickened tissue; 2. short, thick stalk at the base of the lemma in grasses (Poaceae); 3. undifferentiated tissue (parenchyma) often produced in the early stages of plant tissue culture

calyculate

having an epicalyx or similar whorl of bracts subtending the calyx or involucre

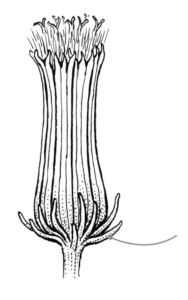

calyculus

bracts subtending the flower and appearing
as a whorl beyond the calyx

SYNONYM epicalyx

calyptra

hood or cap, as with the fused calyx of
California poppies (*Eschscholzia californica*) or
onions (*Allium*)

cambium

lateral meristem that produces cork or
vascular tissue (xylem toward the middle
and phloem toward the outside) and is thus
responsible for the thickening growth of
woody stems and roots

campanulate

shaped like a bell

canaliculate

having one or more longitudinal grooves

SYNONYM channeled

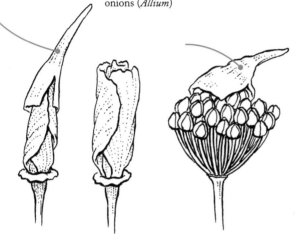

calyx

collective term for flower sepals

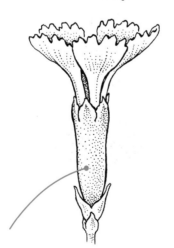

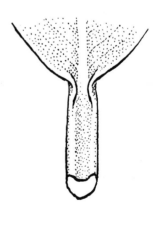

candle

new shoot on a conifer as it's coming out of bud

canopy

1. upper branching portion of a tree; 2. upper layer of a forest consisting mostly of treetops

C

cane

1. stem of a shrub, usually applied to fruit-bearing plants, e.g., raspberries and blackberries (*Rubus*) and roses (*Rosa*); 2. grass stem, usually applied to species that produce large, stiff, or woody stems; 3. long, thin pseudobulbous stem of some orchids, e.g., *Dendrobium*

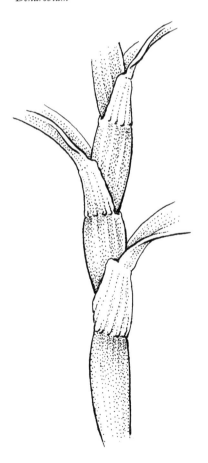

capitate

with a compact head, like a pushpin; commonly used to describe stigmas

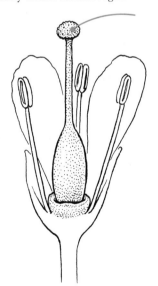

capitulum

(plural capitula) inflorescence of sessile flowers borne on a flattened and expanded portion of the inflorescence axis; the inflorescence of the sunflower family (Asteraceae)

SYNONYM head

capsule

1. dry, dehiscent multilocular fruit splitting open along one to many lines or locations of dehiscence—may be circumscissile, loculicidal, poricidal, septicidal; 2. the spore-containing structure (sporangium) of mosses

carotene

yellow, orange, and red pigments in plants, important in photosynthesis, oil-soluble

carpel

1. primary unit of a pistil, consists of an ovary, style, and stigma and contains ovules. May be solitary in the flower (one simple pistil), or in multiples either free (more than one simple pistil) or fused (compound pistil); 2. megasporophyll of angiosperms

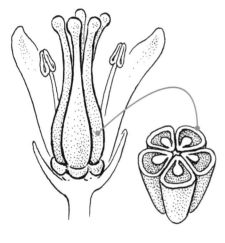

carpellate

having carpels

carpet-forming

densely growing prostrate or short-statured plants covering the ground

carpophore

central stalk-like structure from an extended receptacle to which carpels are attached, as with some species in the carrot (Apiaceae), geranium (Geraniaceae), and buttercup (Ranunculaceae) families

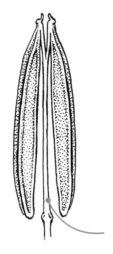

cartilaginous

cartilage-like; tough but flexible tissue

caruncle

outgrowth of the ovule's outer integument, visible as a raised area or appendage on the seed coat near the micropyle and hilum

caruncular

pertaining to caruncles

carunculate

having a caruncle

caryopsis

dry, indehiscent fruit in which the single seed is fused to the pericarp; fruit of the grass family (Poaceae), derived from a unicarpellate pistil

SYNONYM grain

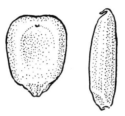

castaneous

reddish brown, rust- or chestnut-colored

SYNONYM ferruginous, rufous, rufus

catkin

usually pendulous spike-like inflorescence of unisexual, sessile to subsessile, apetalous flowers

SYNONYM ament

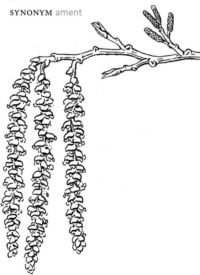

caudate

tapering to a long point, with concave margins at the base of the tip

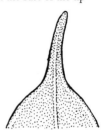

caudex

(plural caudices, caudexes) 1. persistent above- or at-ground portion of herbaceous perennials; 2. main plant axis, the roots and stem; 3. water-storing swollen plant base

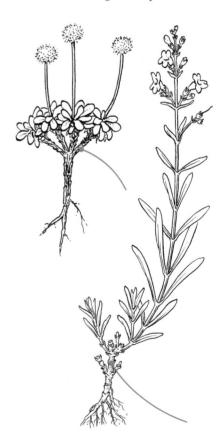

caudiciform

having a water-storing swollen plant base (caudex) from which the stems arise, as with some arid-adapted species in the spurge (Euphorbiaceae), gourd (Cucurbitaceae), and bean (Fabaceae) families

cauliflorous

having flowers and fruits borne along the trunk and/or branches

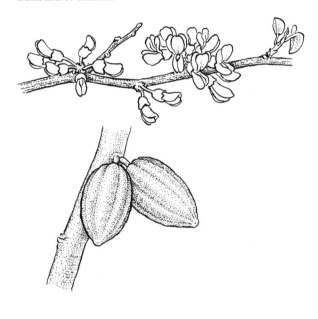

cauline

emerging from or pertaining to the stem, applied most often to leaves

caulescent

clearly having an above-ground stem

ANTONYM acaulescent

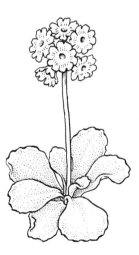

cell

1. cavity or chamber, as with anther sacs and locules; 2. smallest unit of organization in organisms

ceraceous

waxy physically or visually

cernuous

hanging or bent downward, applied most often to flowers

SYNONYM nodding

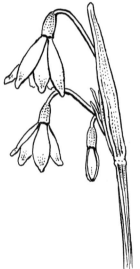

cespitose, caespitose

growing in dense clusters

SYNONYM clumped

chaff

dry, fine bracts or scales, like those subtending the fruits of the sunflower (Asteraceae) and grass (Poaceae) families

chaffy

having chaff

chamaephyte

plant that bears resting or overwintering buds on or near the ground

chambered

having multiple cavities, or hollow sections, separated by walls; e.g., pith of black walnut (*Juglans nigra*)

channeled

having one or more longitudinal grooves

SYNONYM canaliculate

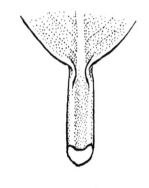

chartaceous

having a papery texture

chasmogamous

flowers that are fertilized after opening (at anthesis), generally cross-pollinated

ANTONYM cleistogamous

chimera, chimaera

plant that contains two different types/strains of DNA, as results from grafting, genetic engineering, or mutation

chiropterophilous

pollinated by bats

chiropterophily

bat pollination

chlorophyll

green pigment in plants, photosynthetic, fat-soluble

chlorophyllous

having chlorophyll

chlorosis

yellowing of tissue due to insufficient chlorophyll production, usually caused by nutrient deficiency

ciliate

with a fringe of hairs extending from the margin

cincinnus

(plural cincinni) ambiguous term that is variously used as a synonym of helicoid cyme or scorpioid cyme

cincturing

removing only a very narrow and thin layer of bark tissue to increase fruit set and size in fruit-bearing plants; e.g., peach (*Prunus persica*) and grapes (*Vitis*)

SYNONYM girdling

circinate

coiled like watch springs, as with the arrangement of fern fronds in bud

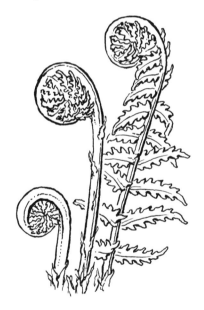

circular

round

SYNONYM orbicular

circumscissile

dehiscing transversely so the top comes off, as with some anthers and capsules; see also loculicidal, poricidal, septicidal

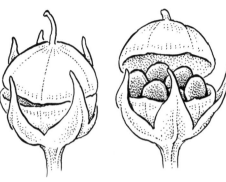

cirrose, cirrhose

having tendrils or terminating in a tendril

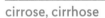

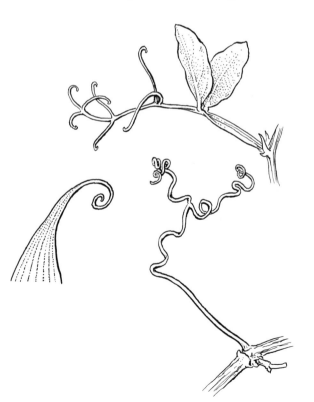

cladophyll, cladode

stem that looks and functions like a leaf
SYNONYM phylloclade

cladoptosic

deciduous by branches and leaves shedding together; e.g., bald cypress (*Taxodium distichum*)

clambering

climbing, but very weakly or not at all attached to supporting structures

clasping

surrounding the stem partially to nearly completely, as with some leaf blade bases in the grass family (Poaceae)

class

taxonomic rank above order and below division

clavate, claviform

shaped like a club

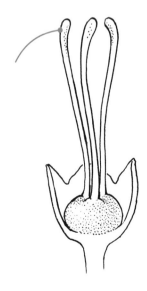

claw

narrowed base of an otherwise broad structure; e.g., some petals and sepals; see also unguiculate

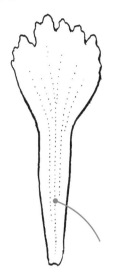

cleft

notch or sinus extending almost to the middle of the structure, usually used in reference to petals or leaves

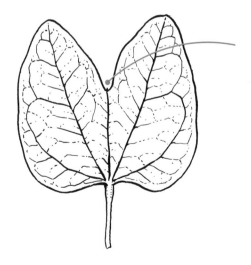

cleistogamous

flowers that are self-fertilized while still closed; e.g., violets (*Viola*)

ANTONYM chasmogamous

climber

plant that grows upward by leaning against or attaching itself to support structures, which may be other plants

climbing

growing upward by leaning against or attaching to support structures

clinal variation

structural and/or genetic differences in plant populations along an environmental gradient such as elevation or moisture

clinandrium

part of the orchid column in which the anther is located

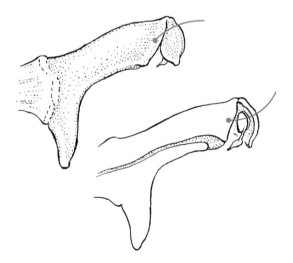

clonal

of or pertaining to a clone

clone

a plant that is genetically identical to one or more other plants, having originated vegetatively from the same parent plant (either naturally or through cultivation)

clove

one section of a bulb; e.g., a segment of garlic (*Allium sativum*)

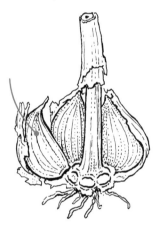

clumped

growing in dense clusters

SYNONYM caespitose, cespitose

coat

covering, as on seeds

coccus

(plural cocci) individual section of a schizocarp, derived from a single carpel in a syncarpous pistil; e.g., cranesbills (*Geranium*)

SYNONYM mericarp

cochleate

forming a spiral like a snail shell

coherent

having like structures weakly stuck together

coleoptile

protective covering on the growing stem tip (plumule) of germinating monocotyledon seeds

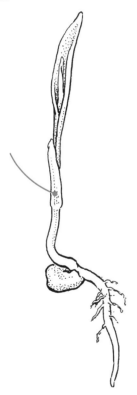

coleorhiza

protective covering on the growing root tip (radicle) of germinating seeds in the grass family (Poaceae)

collar

1. area where a grass leaf's sheath meets its blade; 2. base of a branch where it meets the parent stem and is usually somewhat swollen

columella

central axis of some flowers to which the carpels are attached, persists in fruit

column

1. staminal filaments united into one central structure, e.g., hibiscus (*Hibiscus*); 2. fused filaments and styles, e.g., orchids (Orchidaceae)

columnar

shaped like a column

coma

dense cluster of hairs attached to the end of a seed, facilitates wind dispersal; in milkweeds (*Asclepias*), the coma is sometimes called a pappus

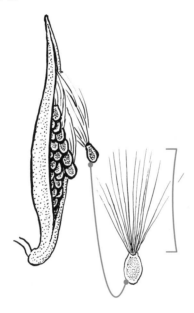

comose

having a dense cluster of hairs (coma)

compatible

1. capable of sexually reproducing together; 2. capable of surviving being grafted together
ANTONYM incompatible

complete

having all parts, as with flowers possessing all four whorls (calyx, corolla, androecium, gynoecium)

composite

vernacular name for any member of the sunflower family (Asteraceae); the name comes from the old, and still accepted, name for the family, Compositae

compound

consisting of more than one section, division, or order of branching, most commonly applied to leaves and inflorescences

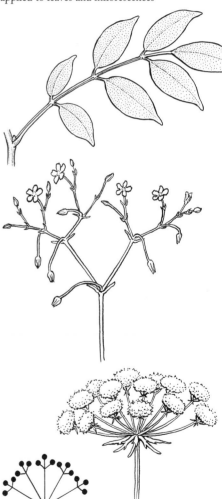

conduplicate

folded from the base to the apex with the upper (adaxial) surface facing itself, as with many fronds in the palm family (Arecaceae)

ANTONYM reduplicate

cone

the reproductive structure of conifers, a central axis with seed- or pollen-bearing sporophylls arranged along it

congested, conglomerate

densely clustered

SYNONYM glomerate

conic, conical

shaped like a cone, attached at the wide end

conifer

plant that has cones and usually needles or scales, evergreen leaves; e.g., pines (*Pinus*) or yews (*Taxus*)

coniferous

having cones

connate

having like structures fused together

ANTONYM discrete, distinct

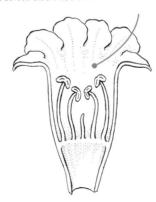

C

connate-perfoliate

opposite leaves, stipules, or bracts fused at the base so appearing to be pierced by the stem

connective

tissue that connects the two anther sacs (thecae) in a stamen

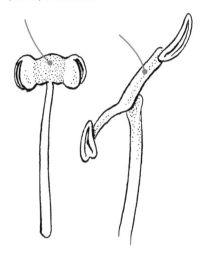

connivent

separated at the base but close at the top, refers to the proximity of two or more unattached structures to each other

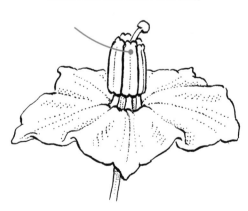

conserved

in nomenclature, a scientific name that has been officially maintained despite some violation of the International Code of Nomenclature for algae, fungi, and plants

conspecific

classified as the same species

constricted

narrowed

SYNONYM contracted

contiguous

touching and continuing without a break but not fused

continuous

unbroken

contorted

twisted, flexed, or bent out of shape

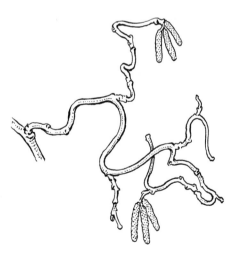

contracted

narrowed

SYNONYM constricted

convolute

like parts arranged so that each one is overlapping the next, as with some flowers' petals

coppice

1. to periodically prune woody plants to the ground for the purpose of causing sprouts to form from the stumps; 2. stand of trees and/ or shrubs that have been cut to the ground to cause sprouting

cordate, cordiform

1. heart-shaped with the widest point at the base; 2. leaf base with lobes shaped like those on a heart

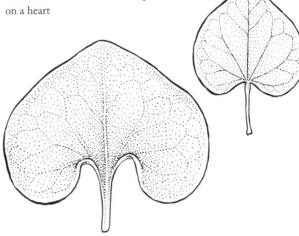

coriaceous

leathery

cork

waterproof outer layer of bark

corm

underground storage structure composed of dense stem tissue covered in papery leaf bases

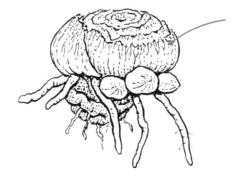

C

cormel

small corm that develops at the base of a larger corm

cornute

horned

corolla

collective term for the petals of a flower

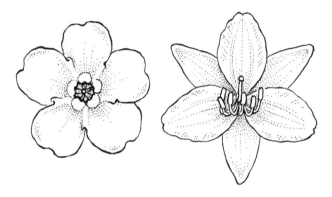

corolla tube

hollow, elongate structure formed by the fusion of a flower's petals

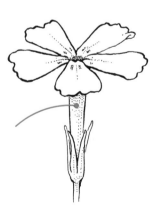

corona

structure or collection of structures between the corolla and androecium of some flowers, often crown-like; e.g., daffodils (*Narcissus*), milkweeds (*Asclepias*)

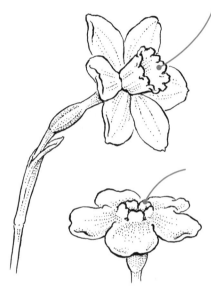

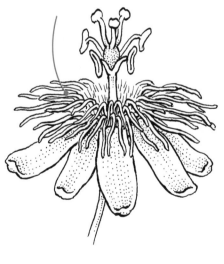

coroniform

shaped like a crown

corpusculum

often sticky central portion of a pollinarium that connects the translator arms that hold the pollinia in milkweeds (*Asclepias*)

corruptule

unfertilized cycad ovule that ripens into what looks like a mature seed but is not viable

cortex

tissue between the vascular tissue and the epidermis in roots and shoots

corymb

branched inflorescence with flowers borne along an elongate axis and having lower branches longer than upper so as to present flowers on a rounded or flat plane at the top

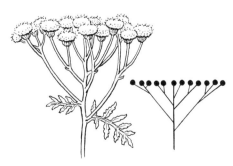

cosmopolitan

distributed worldwide, or nearly so

SYNONYM ubiquitous

costa

(plural costae) leaf or leaflet rib or protruding midvein; in palms (Arecaceae), a petiole extending into the leaf blade

cotyledon

one of a seed's first leaves

SYNONYM seed leaf

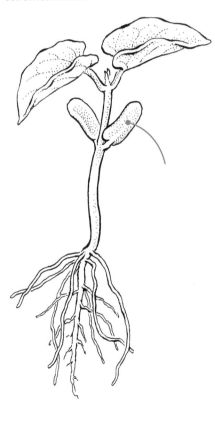

C

creeping
growing along the ground, rooting at the nodes

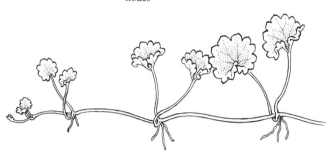

crenate
having a scalloped margin

crenation
individual scallop of a crenate margin

crenulate
having a minutely scalloped margin

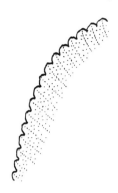

crest
surface ridge

crested
having irregular growth resulting in a mass of tissue produced, usually, at the tip of a stem or inflorescence
SYNONYM fasciated

crispate, crisped
ruffled or crinkled

cross
1. breeding of one organism with another that may be of the same or different species but which possesses different traits; 2. plant resulting from such breeding; see also hybrid

cross-compatible
describes a pair of plants when one is capable of being fertilized by the other and vice versa

cross-pollination

when the pollen of one plant lands and germinates on the stigma of another plant

cross section

cut across the main axis, abbreviated as x.s.

ANTONYM longitudinal section

crotch

axil where two branches or a branch and a trunk join

crown

1. tree apex; 2. top of the persistent portion of an herbaceous perennial; 3. corona

crozier

coiled fern frond in the process of unfurling from bud

SYNONYM fiddlehead

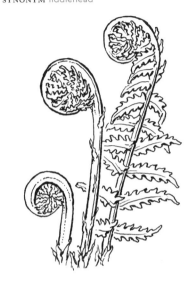

crucifer

vernacular name for any member of the mustard family (Brassicaceae); the name comes from the old, and still accepted, name for the family, Cruciferae

cruciform, cruciate

shaped like a cross

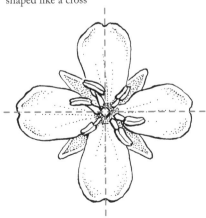

C

crustaceous
brittle

cryptogam
plant reproducing with spores, not seeds
ANTONYM phanerogam

cucullate
hooded

cucullus
hood-shaped structure, especially that of the
corona in milkweeds (*Asclepias*); see also galea
SYNONYM hood

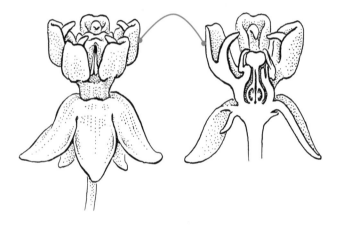

cucurbit
vernacular name for any member of the
gourd family (Cucurbitaceae), including
cucumbers, pumpkins, and squash

culm
jointed stem of grasses and sedges

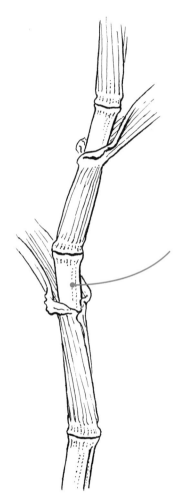

cultigen
a plant that is known only in cultivation, of
which no wild individuals exist

cultivar

a plant selected from the wild or intentionally bred that differs from the typical member of the species from which it was selected or bred; cultivar names must be capitalized and written in single quotes after the scientific name, e.g., *Rhus typhina* 'Tiger Eyes'

cuneate, cuneiform

wedge-shaped with the narrowest point at the base

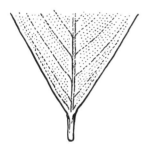

cupulate

shaped like a cup; having a cupule

cupule

cup-shaped structure subtending fruit; e.g., an acorn's cap

cusp

abrupt, sharp point; see also mucro

cuspidate

coming to an abrupt, short, stiff, sharp point

SYNONYM mucronate

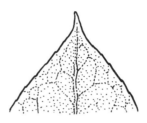

cuticle

waxy outer layer of the epidermis

cutting

a part of a plant usually taken for use in propagation

cyathium

(plural cyathia) "false flower" inflorescence of the genus *Euphorbia*

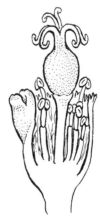

cycads

gymnosperms that resemble palms
(Arecaceae) but which reproduce with cones
(with the exception of female plants of the
genus *Cycas*, whose sporophylls are not
aggregated into cones)

cyclic

occurring in whorls

cylindrical, cylindric

shaped like a cylinder

cyme

unbranched or branched inflorescence with
the axis terminating in the oldest flower that
is basally or centrally located, younger flowers
branch on one or more sides; compound
cymes have flowers presented on a rounded
or flat plane at the top

cymose

having a cyme

cypsela

small, dry, indehiscent fruit derived from
a pistil of two fused carpels; e.g., seeds of
dandelion (*Taraxacum officinale*) and other
fruit in the sunflower family (Asteraceae)

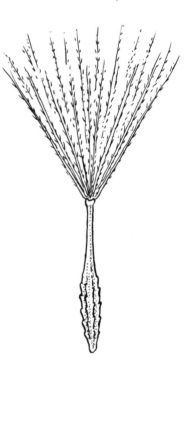

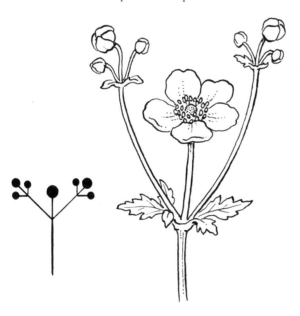

D

damping off

rotting of seedlings caused by a number of pathogens and resulting in death

DBH

= Diameter at Breast Height; measurement of trunk size taken at about one and a half meters (five feet) above ground, the average height of an adult's chest

deadhead

to remove old flowers to encourage more flowers to bloom

deciduous

falling from attachment as with leaves and stipules; see also persistent

ANTONYM evergreen

decumbent

lying or growing flat along the ground with the ends turning upward

decurrent

fused or clasping downward, as with some leaf bases and stipules

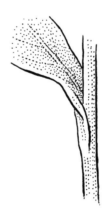

decussate

opposite leaves arranged on the stem at alternating angles of 90° so as to form a cross pattern when viewed down the branch from the tip to the bottom or vice versa

deltoid, deltate

equilateral triangle–shaped, with one of the flat sides on the bottom and the point of attachment in the middle of that flat side

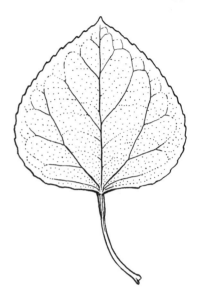

dendriform, dendroid

shaped like a tree

dendritic

branched like a tree

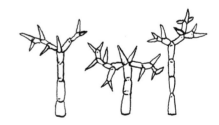

dehiscence

method of opening; e.g., circumscissile, loculicidal, poricidal, septicidal

dehiscent

opening at maturity, as with anthers to release pollen and some fruits to release seeds
ANTONYM indehiscent

dentate

having teeth pointing outward, perpendicular
to the margin

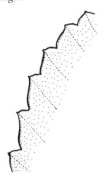

dentation

individual tooth of a dentate margin, or the
whole margin itself

denticulate

having minute teeth pointing outward,
perpendicular to the margin

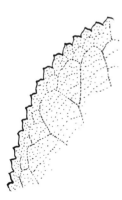

depauperate

fewer than or smaller than normally expected

depressed

pushed in or flattened from above

descending

pointing downward

D

determinate

1. inflorescence of flowers that mature from
the top down or from the center to the sides;
2. growth that terminates in the production
of an organ such as a flower, or the abortion
of the apical meristem

di-

prefix meaning two

diadelphous

stamens fused together by their filaments
into two groups; e.g., pea flowers (Fabaceae
subfamily Papilionoideae)

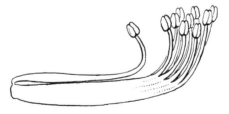

diandrous
having two stamens

dichasium
cymose inflorescence with two opposite flowers (simple) or branching units (compound) produced laterally from each axis

dichotomous
1. branching pattern in which each branch splits into two branches; 2. type of identification key, in which each step has two choices and each choice leads to a different subset of options

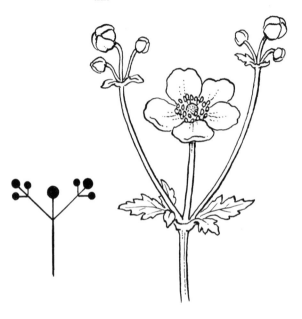

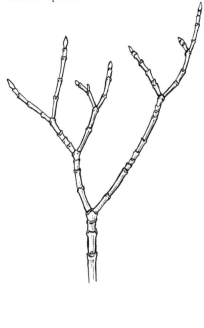

dichogamous
having female and male reproductive whorls maturing at different times
ANTONYM homogamous

dicot
shortened name (from dicotyledon) for the artificial assemblage of plants that typically have two seed leaves (cotyledons), flower parts in multiples of four or five, and net-veined leaves
ANTONYM monocot

dicotyledonous

having two seed leaves (cotyledons)

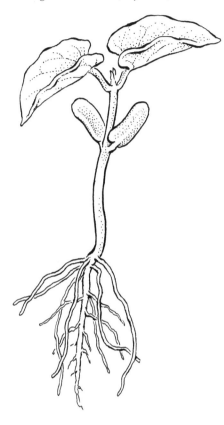

didymous

occurring in pairs

didynamous

having two pairs of stamens that are not
equal in length

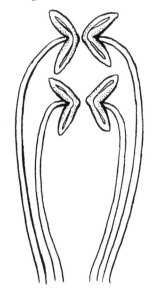

diffuse

spreading

digitate

having veins, lobes, leaflets, or dissections all
arising from a single point, usually at the top
of the petiole, like fingers from a palm

SYNONYM palmate

dilated

expanded

dimorphic

having two different forms, as in having fertile and sterile fronds of two different morphologies; e.g., sensitive fern (*Onoclea sensibilis*)

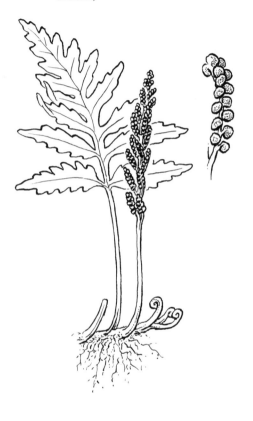

dioecious

having unisexual flowers borne on separate female or male individuals

ANTONYM monoecious

diploid

having two sets of chromosomes (2n); see also haploid, polyploid, tetraploid

diplostemonous

1. having two times the number of stamens as petals; see also haplostemonous; 2. having two distinct sets of stamens, the outer whorl opposite the sepals and the inner whorl opposite the petals

ANTONYM 2. obdiplostemonous

disarticulating

separating along a juncture at maturity

discoid

1. disk-like; 2. inflorescence composed of only disk flowers, as with some plants in the sunflower family (Asteraceae)

discrete

free from similar parts

SYNONYM distinct

ANTONYM connate

disk, disc

expansion of receptacle tissue in some flowers, may be nectariferous

disk flower

flower in the sunflower family (Asteraceae) lacking an expanded corolla and usually occurring in the central portion of the capitulum/head inflorescence

ANTONYM ligulate flower, ray flower

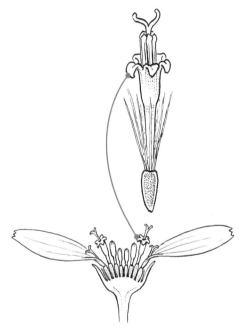

dissected

deeply divided

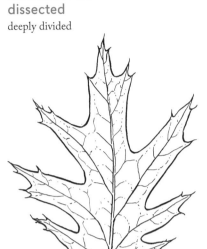

distal

tip, the end farthest from the point of attachment

ANTONYM proximal

distichous

occurring in two opposite rows along a central axis, as with leaves on a stem, making the entire structure appear flat when viewed down the axis from the tip to the bottom or vice versa

SYNONYM two-ranked

distinct

free from similar parts

SYNONYM discrete

ANTONYM connate

divaricate

widely spreading, usually in reference to branching

divergent

spreading

divided

split into two or more segments

division

1. method of propagation in which a perennial plant (or more commonly, a clump of perennial plant clones) is physically split into two or more plants; 2. taxonomic rank above class and below kingdom, plant equivalent to phylum for animals, plant division names end in "-ophyta"

domatium

(plural domatia) small pit or pubescent surface that serves as shelter for invertebrates, often in vein axils

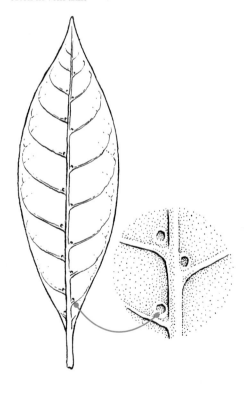

dormant

not actively growing, resting

dorsal

1. pertaining to the back, the surface facing
away from the axis, variously treated as
synonymous with the upper (adaxial) surface
or the lower (abaxial) surface; 2. the upper,
vertical sepal in orchid flowers

ANTONYM 1. ventral

dorsifixed

attached on the back, as with filaments
attached to the back of anthers; see also

basifixed, medifixed, versatile

dorsiventral

1. upper (ventral) and lower (dorsal) surfaces
with different appearances; 2. flattened

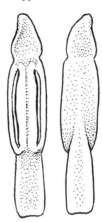

doubled

having more than the usual number of parts
in a floral whorl; e.g., roses (*Rosa*) that have a
proliferation of petals

SYNONYM pleiomerous

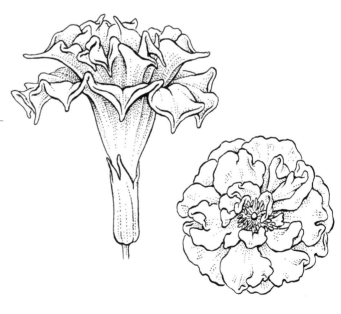

double samara

fruit derived from a two-carpellate ovary that splits at maturity into two winged sections (mericarps that resemble samaras); e.g., the fruit of maples (*Acer*)

SYNONYM samaroid schizocarp

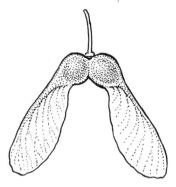

doubly crenate

two-tiered scalloped margin where larger scallops have smaller scallops on them

SYNONYM bicrenate

doubly serrate

margin teeth having teeth of their own, all pointing up toward the apex

SYNONYM biserrate

drip tip

elongated, pointed leaf apex, which allows excess water on the leaf surface to drain quickly

drooping

hanging or bent downward, as with herbaceous parts of dehydrated plants

dropper

shoot growing downward from a bulb or corm, ending in a new bulb or corm

SYNONYM sinker

drupaceous

1. drupe-like; 2. bearing drupes

drupe

fleshy, indehiscent fruit consisting of an exocarp (usually thin), mesocarp (usually fleshy), and a bony or stony endocarp (also called a stone or pit); e.g., peaches and cherries (*Prunus*)

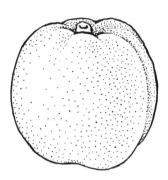
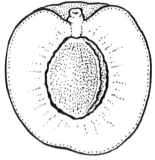

drupelet

small drupe, as with those formed by individual pistils in an aggregate fruit; e.g., raspberries and blackberries (*Rubus*)

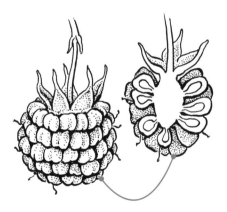

E

e-
prefix meaning lacking, without; see also ex-

ear
grass spike inflorescence and the infructescence that develops from it; e.g., ear of corn (*Zea mays*)

eared
earlobe-shaped; having auricles
SYNONYM auriculate

eccentric
not centrally located on an axis

echinate

having short stiff hairs or prickles

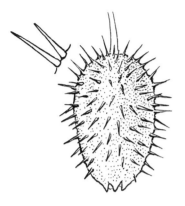

echinulate

having very short stiff hairs or prickles

edaphic

pertaining to soil; used in the context of how soils influence plant growth and communities

eglandular

lacking glands

elaiosome

fleshy appendage near the hilum on the seed coat, attracts ants for seed dispersal, as with the seeds of violets (*Viola*)

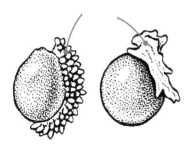

ellipsoid

three-dimensionally ellipse-shaped, broadest in the middle and circular in cross section

elliptic

ellipse-shaped, broadest in middle

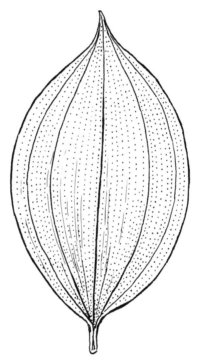
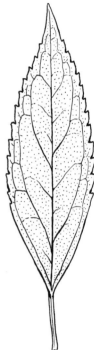

elongate

longer than wide

emarginate

having a rounded apex with an abrupt and shallow indentation in the center

SYNONYM retuse

embryo

immature plant in a seed

emergent

growing above the surface of, e.g., water or the forest canopy

emersed

rising above the water's surface, as with some aquatic plants

ANTONYM submerged, submersed

enation

outgrowth from a surface, as in the leaf-like structures on whisk ferns (*Psilotum*)

SYNONYM excrescence

endemic

native and restricted to a particular area, habitat, or soil type

endocarp

innermost layer of the fruit wall (pericarp) e.g., the pits of peaches and cherries (*Prunus*)

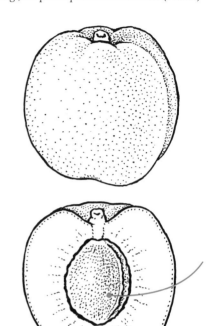

endosperm

seed nutritive tissue for the developing embryo

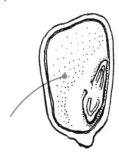

ensiform

lance- or sword-shaped with the widest point
toward the base

SYNONYM gladiate, lanceolate

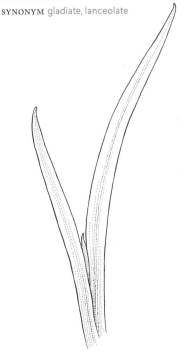

entire

undivided and without lobes or teeth along
the margin

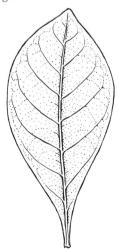

entomophagous

insect-eating

SYNONYM insectivorous

entomophilous

pollinated by insects

ephemeral

short-lived or short-lasting; most commonly
applied to spring ephemerals, which are
plants that grow, flower, fruit, and die back
completely by mid summer

SYNONYM evanescent

epi-

prefix meaning on (e.g., epipetalous) or above
(e.g., epigeal)

epicalyx

bracts subtending the flower and appearing
as a whorl beyond the calyx

SYNONYM calyculus

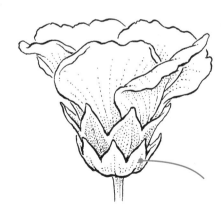

epicarp

outermost layer of the fruit wall (pericarp);
e.g., the skin of a peach (*Prunus persica*)

SYNONYM exocarp

epicotyl

the region of an embryo or seedling above
the cotyledons

ANTONYM hypocotyl

epidermal

pertaining to the epidermis

epidermis

multi-layered outermost surface tissue of
plants

epigeal, epigeous

germination type in which the cotyledons
rise above ground attached to the developing
seedling and become photosynthetic

ANTONYM hypogeal, hypogeous

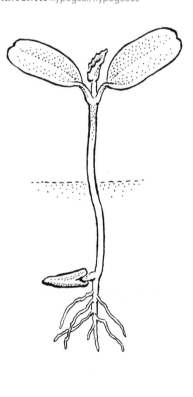

epigynous
flower with an inferior ovary

epilithic
growing attached to rock
SYNONYM epipetric

epipetalous
attached to petals

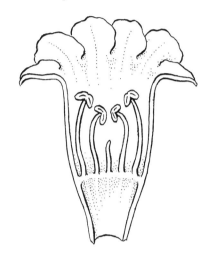

epipetric
growing attached to rock
SYNONYM epilithic

epiphyllous
growing attached to the leaf of another plant but not parasitizing that plant

epiphyte
a plant that grows attached to another plant but which does not parasitize that plant

epiphytic
growing attached to another plant but not parasitizing that plant

epizoochory

seeds externally animal-dispersed, as with
seeds or fruits that stick to animal fur

equilateral

having equal sides

equitant

having partially concentric leaf bases, as with
irises (*Iris*)

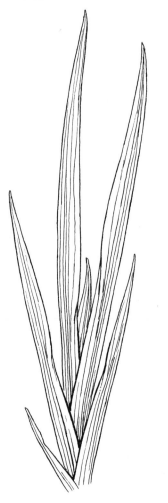

erect

vertical, upright

erose

irregularly toothed

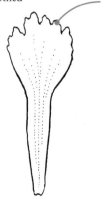

escaped

said of a plant that is now reproducing on its
own in the wild but which arrived in the area
through cultivation

espalier

1. the method of training trees or shrubs to grow flat against a wall or fence, or in the form of a wall; 2. a plant that has been grown in this way

estipellate

lacking stipels

SYNONYM exstipellate

estipulate

lacking stipules

SYNONYM exstipulate

estivation, aestivation

arrangement of perianth parts in bud; see also vernation

etaerio

formed from the fusion of multiple, separate unicarpellate pistils in a single flower, may consist of tiny versions of one of many different fruit types including samaras, drupes, achenes, follicles, etc.; e.g., raspberries and blackberries (*Rubus*)

SYNONYM aggregate fruit

etiolated

pale and elongated growth due to lack of sunlight

evanescent

short-lived or short-lasting; most commonly applied to spring ephemerals, which are plants that grow, flower, fruit, and die back completely by mid summer

SYNONYM ephemeral

even-pinnate

pinnately compound with an even number of leaflets, terminating in a pair of leaflets; see also imparipinnate, odd-pinnate

SYNONYM paripinnate

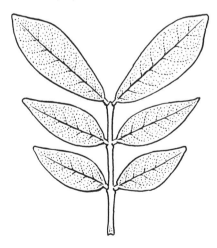

E

everbearing

continuously producing flowers and fruits
throughout the growing season

evergreen

holding on to at least some live leaves
throughout the year
ANTONYM deciduous

everlasting

flower that very closely resembles its fresh
state when dry; e.g., strawflower (*Xerochrysum
bracteatum*) and other members of the
sunflower family (Asteraceae)

ex-

prefix meaning lacking, without; see also e-

excrescence

outgrowth from a surface, as in the leaf-like
structures on whisk fern (*Psilotum nudum*)
SYNONYM enation

exfoliate

remove the outer layers in thin pieces, as with
the bark of birches (*Betula*)

exine

outermost layer of the pollen grain wall

exocarp

outermost layer of the fruit wall (pericarp);
e.g., the skin of a peach (*Prunus persica*)
SYNONYM epicarp

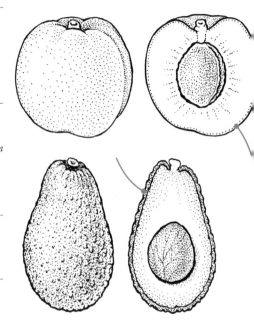

exotic

not native to a particular area, habitat, or soil
type; introduced; e.g., tropical plants grown
in temperate climates

explant

in tissue culture, the part of the parent plant
that is transferred to growth medium for
culturing

exserted

projecting or protruding, as with a style
extending beyond a corolla

ANTONYM included

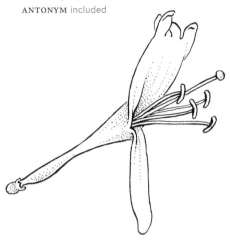

ex situ

in a created environment, in cultivation

exstipellate

lacking stipels

SYNONYM estipellate

exstipulate

lacking stipules

SYNONYM estipulate

extra-

prefix meaning outside

extrafloral

outside of the flower, as with nectaries
located on leaves

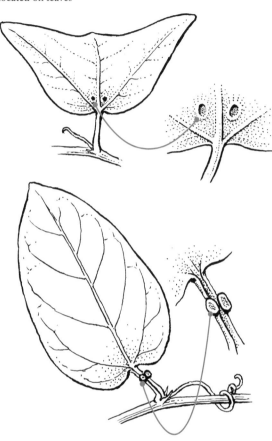

extrastaminal

outside of the staminal whorl

extrorse

stamens facing and dehiscing out and away
from the flower's center

exudate
liquid released from damaged tissue

eye
1. node of some tubers, such as potato (*Solanum tuberosum*); 2. subterranean young vegetative or reproductive shoot buds of dahlias (*Dahlia*)

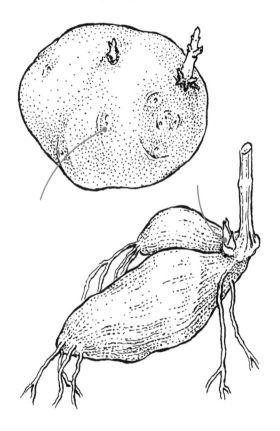

face
inner or upper surface of a plant organ

falcate, falciform
curved to one side like a sickle

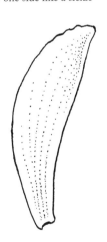

fall
one of the three outer tepals (all sepals) in the flowers of irises (*Iris*); see also standard

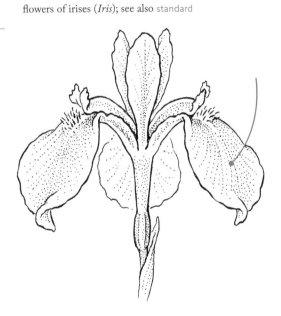

fall-bearing
a fruiting shrub with primocanes that produce fruit in the autumn of their first year of growth; e.g., some raspberries and blackberries (*Rubus*); see also summer-bearing

false flower

inflorescence that closely resembles a single
flower; e.g., dogwood (*Cornus florida*),
Euphorbia cyathia, Asteraceae heads/capitula
SYNONYM pseudanthium

false fruit

seed-bearing structure resembling and
often mistaken for a fruit but for which the
majority of the tissue is non-ovary (may be
from such structures as a hypanthium or
receptacle); e.g., rose hips (*Rosa*)
SYNONYM anthocarp, pseudocarp

false indusium

(plural false indusia) a pocket in the margin
of a frond or fold of the frond that covers a
fern sorus

family

taxonomic rank above genus and below
order; plant family names end in "-aceae"

fasciated

having irregular growth resulting in a mass of
tissue produced, usually, at the tip of a stem
or inflorescence
SYNONYM crested

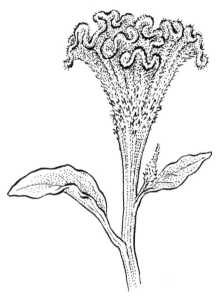

fascicle

a bundle of like organs; e.g., pine needles

fasciculated

occurring in bundles or clusters

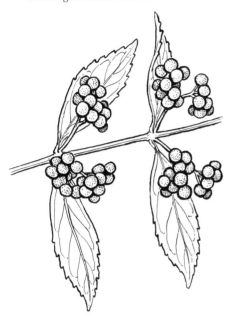

fastigiate

having branches that occur in an erect broom-like cluster

faveolate, favose

honeycomb-like with neatly arranged depressions and ridges

SYNONYM alveolate

feather

lateral branch on a current year's stem

female flower

flower bearing fertile female structures (pistils) and no, or only infertile, male structures (stamens)

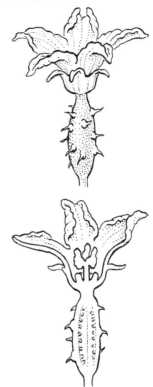

fenestrate

having small window-like areas

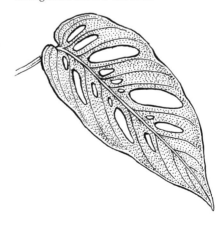

ferruginous

reddish brown, rust- or chestnut-colored

SYNONYM castaneous, rufous, rufus

fertile

1. capable of sexual reproduction, may refer to an individual part (e.g., flower, frond, pistil, stamen) or an entire individual; 2. bearing flowers, cones, spores, or seeds

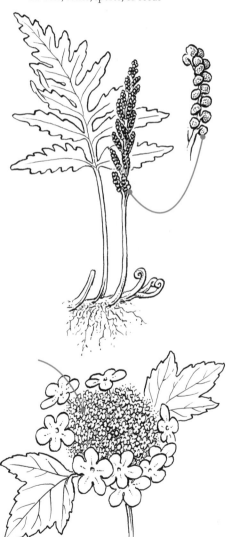

fetid

having a bad odor

fibrous

1. having fibers; 2. fiber-like

fibrous roots

root system in which the roots are of roughly the same diameter

fiddlehead

coiled fern frond in the process of unfurling from bud

SYNONYM crozier

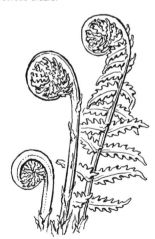

filament

1. stalk bearing the anther in a stamen; 2. thin fiber

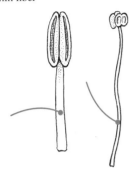

filamentous

1. having filaments; 2. filament-like

filiform

thread-like

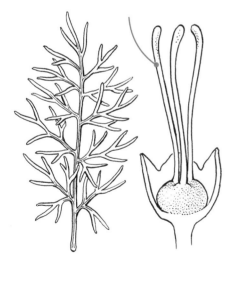

fimbriate

fringed in hairs, applied to margins

first leaf

the first leaf of a seedling after the seed leaves, often markedly different in morphology from the mature plant's leaves

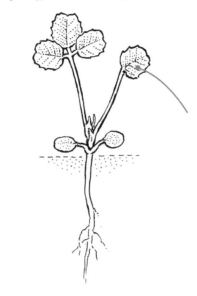

flabellate, flabelliform

shaped like a fan; e.g., ginkgo (*Ginkgo*) leaf

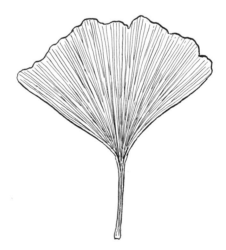

flagellate

having long, thin runners (stolons)

SYNONYM sarmentose

floral cup

tubular structure surrounding and fused to
or free from the ovary, may be an expansion
of the receptacle and/or a fusion of various
components of the outer three floral whorls
(calyx, corolla, androecium)

SYNONYM hypanthium

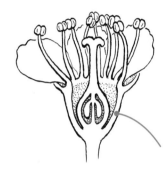

fleshy

succulent, water-conserving tissue

flexuose, flexuous

bending back and forth at angles in opposite
directions, zigzagging, as with stems that
have sympodial growth

floral envelope

collective term for calyx (sepals) and corolla
(petals)

SYNONYM perianth

floral tube

fused tubular calyx or corolla

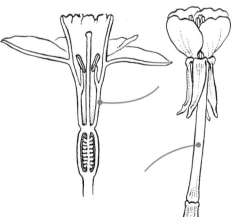

floral

pertaining to flowers

floret

1. small flower; 2. single flower within an inflorescence, e.g., in the sunflower (Asteraceae), carrot (Apiaceae), and mustard (Brassicaceae) families; 3. smallest unit of the grass family (Poaceae) inflorescence, consisting of a flower and the two subtending bracts (lemma and palea)

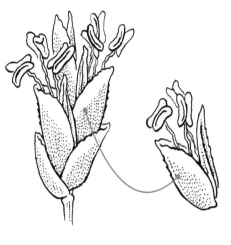

floricane

a fruiting shrub's second-year stem that bears fruits midway through the growing season; in some plants, such as raspberries and blackberries (*Rubus*), individuals with floricanes are called summer-bearing

ANTONYM primocane

flower

organ in angiosperms that, when complete, bears the female (pistil) and male (stamen) reproductive structures as well as a calyx and corolla

flush

emergence of leaves or flowers on woody plants

fluted

grooved with regularity in spacing, applied to cylindrical structures

foliaceous

leaf-like, usually applied to a bract or sepal

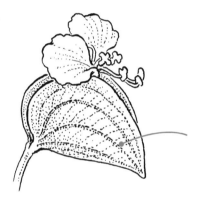

foliage

collective term for a plant's leaves

foliar

1. relating to leaves; 2. leaf-like

foliate

bearing leaves

foliolate

1. bearing leaflets; 2. leaflet-like

F

follicle

dry, dehiscent unilocular fruit opening along one line of suture, derived from a unicarpellate pistil; e.g., milkweeds (*Asclepias*)

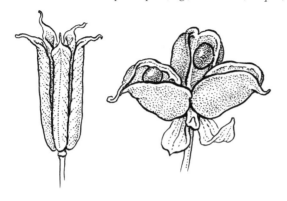

fornix

(plural fornices) small arched projection in a flower's throat (inside a tubular corolla), as in many members of the borage family (Boraginaceae)

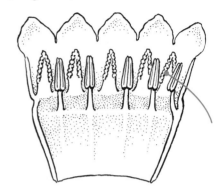

forb

broadleaf herbaceous, non-tree-like plant, often applied to non-grass-like herbaceous plants that are grazed by animals

force

to cause a plant to bloom out of its normal season or sequence by using horticultural techniques

form, forma

taxonomic rank below species; individuals or populations usually differ from what is typical for the species in only very minor ways compared with those characters that define subspecies or varieties

foveolate

pitted with small depressions

free

not fused

free-central placentation

ovules borne on the freestanding central column of a unilocular ovary

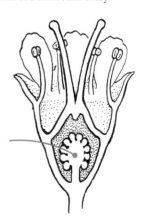

frond

the leaf of a fern, palm, or cycad

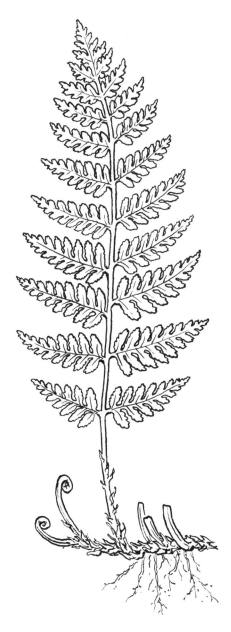

frost heaving

phenomenon in which plants and soil move
as a result of water freezing

SYNONYM heaving

fruit

sexual reproductive structure in which seeds
are housed, a ripened ovary

fruit set

very early stage of fruit development when
the ovary is just beginning to transform into
the fruit; often marked by the style, corolla,
and androecium falling off of the flower and
the ovary starting to swell slightly

fruticose, frutescent

shrub-like

fulvous

reddish brownish yellow

funicle, funiculus

stalk that connects an ovule to an ovary wall
or a seed to a fruit wall

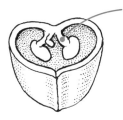

funnel-shaped, funnel-form
shaped like a funnel with sides tapering
downward into a narrow cylinder

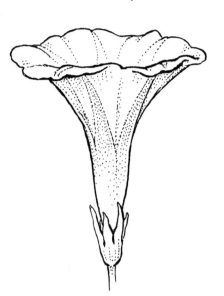

furfuraceous
covered in soft, flaky scales

furrowed
having deep longitudinal grooves, most often
applied to bark

fused
attached; includes both adnate and connate

fusiform
three-dimensionally shaped like a spindle,
widest at the middle and tapering to a point
at both ends

galea

upper petal or other floral structure that is helmet- or hood-like; e.g., monkshood (*Aconitum*)

gall

mass of plant tissue developed around a wound from a parasitic insect, mite, bacterium, fungus, etc.

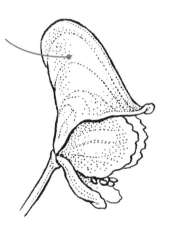

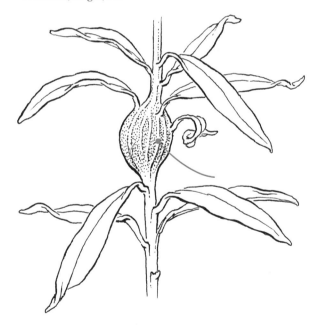

gametes

cells that combine in sexual reproduction, egg and sperm

gametophyte

life cycle generation in which a plant has one set of chromosomes (i.e., is haploid, 1n) and produces gametes (sperm or eggs); in seed plants the gametophyte is the ovule and the pollen grain; the gametophyte is dominant in both time and size for non-vascular plants, making it their most conspicuous generation, as with mosses

ANTONYM sporophyte

gamo-

prefix meaning having like structures fused together (connate)

gamopetalous

having a corolla that is at least partially fused

SYNONYM sympetalous

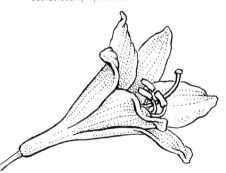

gamosepalous

having a calyx that is at least partially fused

SYNONYM synsepalous

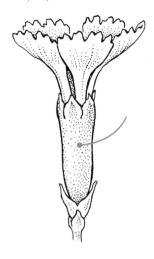

gemma

(plural gemmae) vegetative propagule that develops on a plant as a clump of cells or bud-like structure and then separates from the plant, most often associated with liverworts

genet

a clonal, genetically identical colony, individual plants in the colony are called ramets

geniculate

bent like an elbow

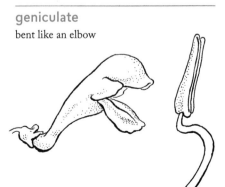

genus

(plural genera) taxonomic rank above species and below family, may also be within a subfamily and/or a tribe

geophyte

plant that survives harsh conditions underground as a root, bulb, corm, rhizome, or tuber

geotropism

growth of roots in the direction of gravity and shoots in the opposite direction
SYNONYM gravitropism

germination

process by which seeds or spores begin to grow

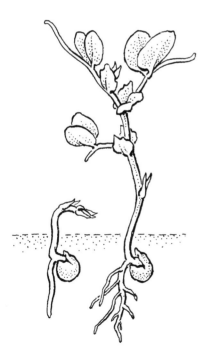

gibbous

bulging on one side

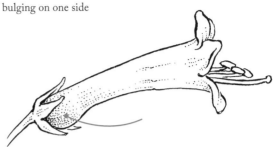

girdling

1. removing or cutting through the bark, including all live vascular tissue, all the way around a woody stem to prevent the flow of water and nutrients, results in the death of the plant or stem above the girdle; 2. removing only a very narrow and thin layer of bark tissue to increase fruit set and size in fruit-bearing plants, e.g., peach (*Prunus persica*) and grapes (*Vitis*)
SYNONYM 2. cincturing

glabrate, glabrescent

becoming hairless (glabrous), as with some leaves as they mature

glabrous

lacking hair

gladiate

lance- or sword-shaped with the widest point toward the base
SYNONYM ensiform, lanceolate

gland

structure that produces oily or sugary secretions, usually to attract insects

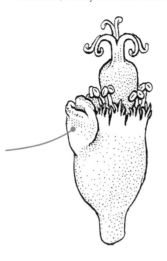

glandular

having glands

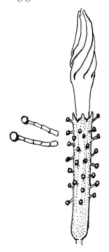

glaucous

having a whitish waxy covering on the surface that easily wipes away

globose, globular

round in three dimensions

SYNONYM spherical

glochid

(plural glochidia) small barbed hair, as found at the nodes of cacti (Cactaceae)

glomerate

densely clustered

SYNONYM congested, conglomerate

glume

one of two lowermost bracts of a spikelet in grasses (Poaceae)

grafting

method of propagation in which two or more woody plants are joined at cut surfaces; there are many types of grafting, but the most common joins the terminal end of a stem (called the scion) with a branch or young trunk from which the terminal portion has been removed (called the rootstock); grafting is the primary method by which woody fruit-bearing plants, e.g., apples (*Malus*), peaches and cherries (*Prunus*), are propagated

grain

1. dry, indehiscent fruit in which the single seed is fused to the pericarp; fruit of the grass family (Poaceae), derived from a unicarpellate pistil; 2. the vertical pattern of fibers in wood

SYNONYM 1. caryopsis

granular

having or consisting of small particles or projections similar to grains

gravitropism

growth of roots in the direction of gravity and shoots in the opposite direction

SYNONYM geotropism

grex

(plural greges) all offspring of a particular intentional hybrid cross, frequent with orchids (Orchidaceae) and azaleas (*Rhododendron*), e.g.; often used as an informal taxonomic rank

ground cover

plants grown for their ability to conceal the soil from view and protect it from erosion, these may be simply short, upright, tightly clumping plants or those that grow along the ground

growth habit

the form or appearance of a plant; e.g., shrub, tree, prostrate, climbing

SYNONYM habit

guttation

expelling of droplets of liquid from the leaf margin and apex

gymnosperm

plant bearing ovules naked upon sporophylls (such as pine cone scales) that develop into seeds upon sporophylls

gynandrium

fused androecium and gynoecium, as in orchids (Orchidaceae)

gynandrous

having male reproductive structures (stamens) fused with female reproductive structures (pistils)

gynobase

enlarged receptacle subtending the gynoecium, as in the mint (Lamiaceae) and borage (Boraginaceae) families

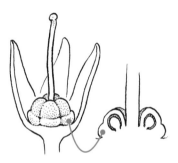

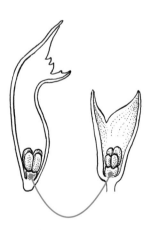

gynoecium

female reproductive portion of the flower, consists of single or multiple pistils

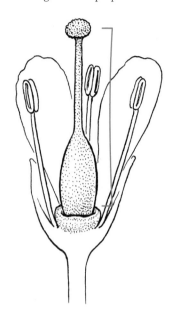

gynophore

stalk elevating the gynoecium

gynostegium

column of fused male (stamens) and female (pistils) reproductive structures, as in milkweeds (*Asclepias*)

habit

the form or appearance of a plant; e.g., shrub, tree, prostrate, climbing

SYNONYM growth habit

habitat

conditions or type of location in which a plant grows; e.g., arid, wet, desert, prairie

haft

1. very narrowed petal or sepal base (claw) of some flowers; 2. petiole or stem with a green wing

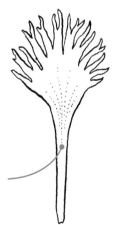

hair

an outgrowth of the epidermis consisting of one or more elongated cells; see also trichome

halophyte

plant that can survive in salty habitats

haploid

having one set of chromosomes (1n); see also diploid, polyploid, tetraploid

haplostemonous

1. having as many stamens as petals; 2. having one set of stamens; see also diplostemonous

hardiness

ability of a plant to survive in the average growing conditions of a particular location, most often used in reference to cold hardiness

hardiness zones

geographic classification system that informs people which plants may survive in particular areas based on their ability to tolerate the average minimum temperature of that area; first developed by the U.S. Department of Agriculture

hastate

arrowhead-shaped with basal lobes pointing outward from the midvein

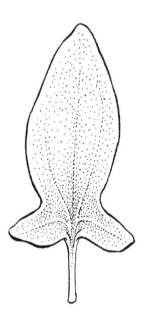

head

inflorescence of sessile flowers borne on a flattened and expanded portion of the inflorescence axis; the inflorescence of the sunflower family (Asteraceae)

SYNONYM capitulum

heartwood

central, older, darker wood of trees that is prized in woodworking

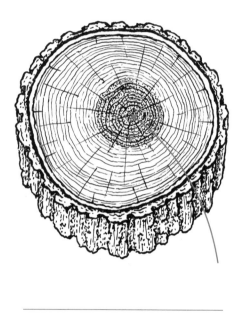

heaving

phenomenon in which plants and soil move as a result of water freezing

SYNONYM frost heaving

helicoid cyme

sympodial inflorescence with flowers borne on one side of the axis and forming a spiral; can be difficult to distinguish from a scorpioid cyme

herbaceous

of or pertaining to plants without above-ground woody growth

herbarium

a natural history collection of dried or otherwise preserved plant specimens

hermaphrodite

plant with bisexual flowers

hesperidium

multilocular berry with leathery exocarp, locules separating as segments in fruit; e.g., the fruit of lemons and oranges (*Citrus*)

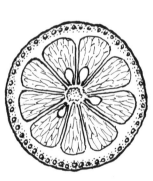

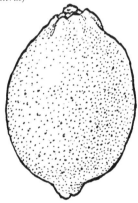

hemi-epiphyte

plant that grows attached to another plant, unrooted to the ground at one point in its life span but rooted to the ground at another; may start as an epiphyte and later become rooted, or start rooted and later become an epiphyte (less common)

herb

1. plant without above-ground woody growth; 2. plant used as a culinary seasoning, food, fragrance, or medicine

hetero-

prefix meaning different

heterogamous

having separate female and male flowers

ANTONYM homogamous

heterogonous

having two or more forms of bisexual flowers on separate individuals that differ in the ratio of the length of stamens to pistil

ANTONYM homogonous

heteromerous

having different numbers of parts, as in flowers with five petals and 10 stamens

heterophyllus

having different types of leaves on one individual

heterosporous

having two different kinds of spores, as is the case with all seed plants, certain aquatic ferns (e.g., *Azolla*, *Marselia*, and *Salvinia*), and two lycophytes (*Isoetes* and *Selaginella*)

ANTONYM homosporous

hilum

scar left by the funicle on the seed coat

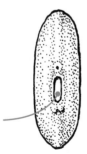

hip

false fruit consisting of a thickened and hardened hypanthium containing achenes, the fruit of roses (*Rosa*)

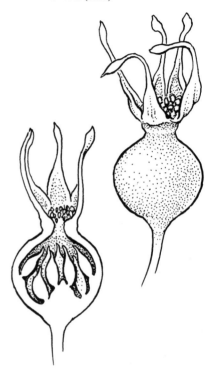

hirsute

having stiff, rough hairs

homo-

prefix meaning same

homogamous

1. having bisexual (perfect) flowers; 2. having female and male reproductive whorls maturing at the same time

ANTONYM 1. heterogamous; 2. dichogamous

homogonous

having only one form of bisexual flowers on different individuals and no difference in the ratio of the length of stamens to pistils
ANTONYM heterogonous

homosporous

having one kind of spore, as in some ferns (pteridophytes)
ANTONYM heterosporous

hood

hood-shaped structure, especially that of the corona in milkweeds (*Asclepias*); see also galea
SYNONYM cucullus

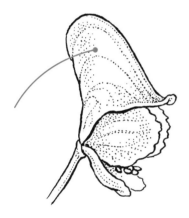

hook

narrow structure or extension that is abruptly curved at the tip

horn

more or less basally cylindrical structure that curves and tapers to a point like a bull's horn

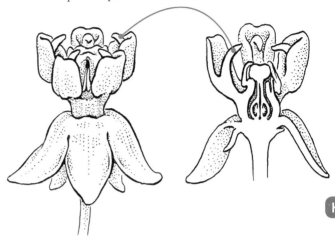

host

plant which another plant parasitizes, extracting nutrients

humus

decayed organic matter

husk

outer layer or layers of seeds or fruits (may correspond to all or parts of the pericarp or seed coat)

hybrid

plant produced via sexual reproduction involving two different species or varieties

hybrid swarm

hybrid plants that readily breed such that original parent plants cross with each other and with hybrids and hybrids cross with each other

hydrophilous

pollinated by water

hydrophyte

plant adapted to growing in water; see also
mesophyte, xerophyte

hygroscopic

readily absorbing moisture from the air

hypanthium

tubular structure surrounding and fused to
or free from the ovary, may be an expansion
of the receptacle and/or a fusion of various
components of the outer three floral whorls
(calyx, corolla, androecium)

SYNONYM floral cup

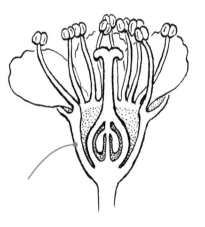

hyphae

"vegetative" strand (branching filament) of a
fungus

hypo-

prefix meaning low or beneath

hypocotyl

the region of an embryo or seedling below
the cotyledons

ANTONYM epicotyl

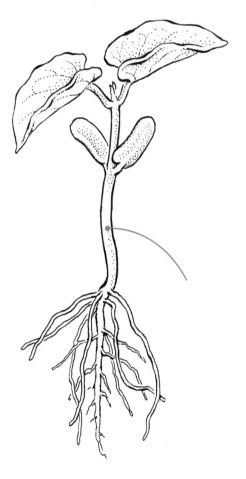

hypogeal, hypogeous

germination type in which the cotyledons
remain below ground, where the seed was
originally located, and do not become
photosynthetic

ANTONYM epigeal, epigeous

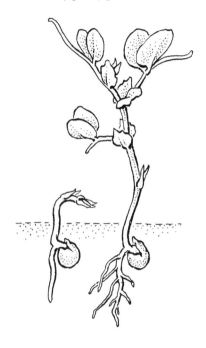

hypogynous

flower with a superior ovary and no floral cup
(hypanthium)

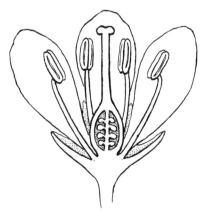

I

ICN

= International Code of Nomenclature for algae, fungi, and plants, dictates the rules for naming all naturally occurring plants (i.e., does not cover cultivars); formerly known as the ICBN (International Code of Botanical Nomenclature)

ICNCP

= International Code of Nomenclature for Cultivated Plants, dictates the rules for naming cultivated plants that are not covered by the ICN; covers cultivars and other human-selected or -bred plants

imbricate

having parts that overlap like roof shingles, commonly used in aestivation to refer to the arrangement of the petals in bud

immersed

submerged

imparipinnate

pinnately compound with an odd number
of leaflets, terminating in a single leaflet; see
also even-pinnate, paripinnate

SYNONYM odd-pinnate

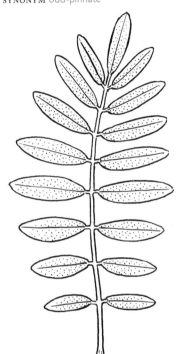

incised

cut angularly, leaving jagged sections

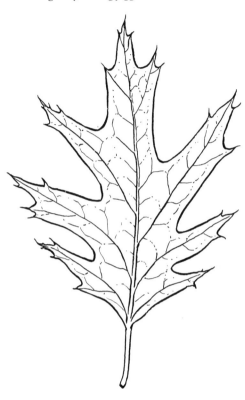

imperfect

having only male (left) or female (right)
functional reproductive parts

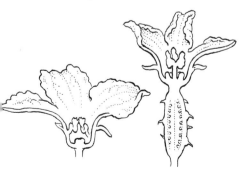

included

not protruding, as with flower parts within
a corolla

ANTONYM exserted

incompatible

1. not capable of sexually reproducing
together; 2. not capable of surviving being
grafted together

ANTONYM compatible

incomplete

missing one or more floral whorls

incurved

curved inward toward the middle or axis

indehiscent

not opening, as with some fruit
ANTONYM dehiscent

indeterminate

1. inflorescence of flowers that mature
from the bottom up or from the sides to
the middle and which has the capacity
for indefinite growth; 2. shoots for which
elongating growth has the capacity to go on
indefinitely

indigenous

originating from a particular geographical or
geological area
SYNONYM native

indumentum, indument

covering of hairs and/or scales on the
epidermis of a plant

indusium

(plural indusia) a flap of tissue that covers a
sorus on a fern frond

inferior

located below, as with an ovary located below
the point of attachment of the outer three
floral whorls (calyx, corolla, androecium)

infertile

not capable of sexual reproduction
SYNONYM sterile

inflated

swollen, distended

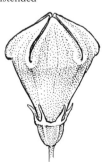

inflorescence

branched or unbranched axis upon which flowers are arranged

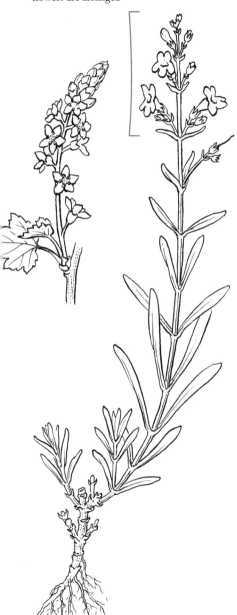

infra-

prefix meaning below

infructescence

branched or unbranched axis upon which fruits are arranged

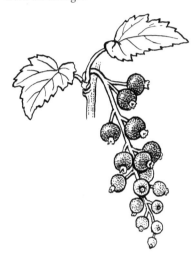

innovation

shoot that may eventually separate from the parent plant and continue to live; e.g., the plantlet produced at the end of a stolon

inosculation

fusing together of woody stems where they come in contact with each other; this can happen within one individual or between two or more individuals

I

inrolled

rolled upward toward the upper (adaxial)
surface

SYNONYM involute

ANTONYM revolute

insectivorous

insect-eating

SYNONYM entomophagous

inserted

fused to or emerging from another structure

in situ

in a natural environment, in the "wild"

integument

tissue surrounding the ovule that becomes
the seed coat as the ovule develops into the
seed

inter-

prefix meaning between

interfertile

said of two or more taxa that may
successfully sexually reproduce with each
other

intergeneric hybrid

offspring produced by the crossing of
different genera, as is the case with many
orchid hybrids

internode

area of stem occurring between the two
nearest locations of leaf attachment (nodes)

ANTONYM node

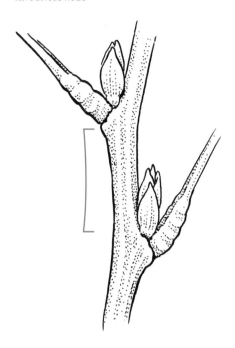

interpetiolar

between petioles

interrupted

discontinuous in structure or content

interspecific hybrid

offspring produced by the crossing of
different species

intra-

prefix meaning within

intrastaminal

borne between the stamens and the gynoecium or center of the flower

introduced

non-native plant intentionally or unintentionally brought into an area; e.g., aquatic plants released into new areas in ballast water

introrse

facing or opening inward toward the center, as with anthers in a flower

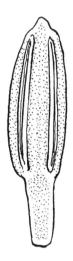

invasive

non-native plant that is reproducing on its own and interfering with the normal functions and/or composition of an ecosystem

inverted

occurring in an orientation opposite to what is normal

involucre

bracts subtending a flower or a collection of flowers, as with the inflorescences of the sunflower family (Asteraceae)

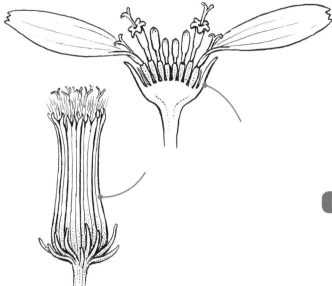

involute

rolled upward toward the upper (adaxial) surface

SYNONYM inrolled

ANTONYM revolute

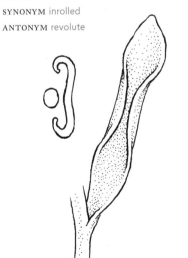

irregular

having a single plane of symmetry such that
only one line drawn through the middle
produces two mirror-image halves

SYNONYM bilaterally symmetrical, zygomorphic

ANTONYM actinomorphic, radially symmetrical,
regular

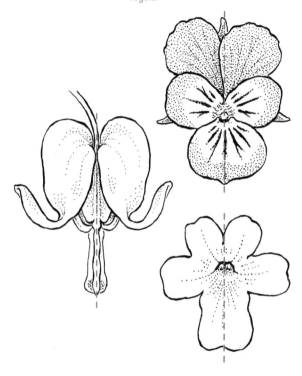

isolation

state of separation, whether temporal or
spatial, to keep plants from breeding

isomerous

having an equal number of parts in the floral
whorls

joint

1. articulation; 2. node, especially in grasses (Poaceae)

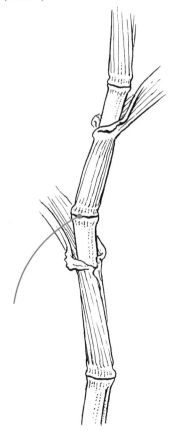

jointed

having nodes that are or appear to be articulated

jugate

having parts in pairs

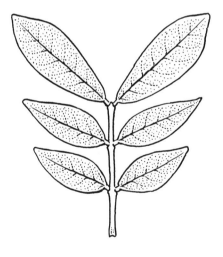

juvenile

plant not yet capable of sexual reproduction, usually smaller in size than adults

K

karyotype

the number, size, and shape of an individual's chromosomes

keel

1. lower central, partially fused two petals in flowers of the bean family (Fabaceae); 2. ridge emerging from a rounded surface

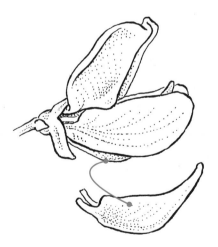

keiki

plantlet produced by an orchid, usually on a long pseudobulbous stem, at the base of a pseudobulb, or on an old inflorescence

key

plant identification tool that allows the reader to select characteristics by which to determine the identity of plants; types include dichotomous, polytomous, multi-entry, and multi-access

knee

1. emergent vertically growing roots of bald cypress (*Taxodium distichum*); 2. bent roots or pneumatophores of mangroves (*Avicennia, Rhizophora*)

labelliform

shaped like a lip

labellum

central, usually lowest and largest petal of an
orchid, may be cup-like

SYNONYM lip

labiate

lip-like or having lips, as with the flowers of
the mint family (Lamiaceae)

labium

(plural labia) lower prominent petal segment of a bilabiate flower, as in the mint family (Lamiaceae)

SYNONYM lip

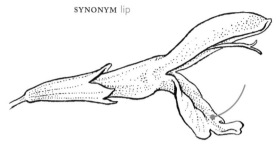

lamina

the usually broad and flattened part of a leaf or petal

SYNONYM blade

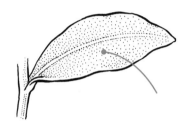

lacerate

irregularly lobed, appearing torn

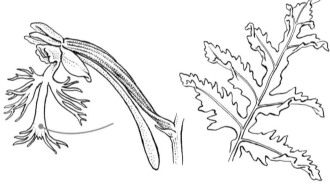

lanceolate

lance- or sword-shaped with the widest point toward the base

SYNONYM ensiform, gladiate

laciniate

deeply lobed into narrow sections

lactiferous, laticiferous

having milky latex

lateral

on or at the side, as with the leaflets below the terminal leaflet on a pinnately compound leaf

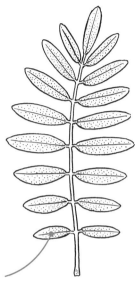

latex

milky sap

latitudinal

perpendicular to the main axis

SYNONYM transverse

latrorse

opening longitudinally on the side, in reference to anther dehiscence

lax

loose, not congested

layering

propagation technique that triggers root production on stems that are still attached to their parent plant; done via several techniques such as putting lower branches partially underground (one to many times, alternating with above-ground sections), or by cutting into stem bark and sealing sphagnum moss or another sterile substrate around the stem at the cut site; the rooted stems are then cut from the plant and resulting new plants are clones of the parent plant

leader

the dominant stem of a tree, the main trunk

leaf

primary photosynthetic organ of most plants, usually attached to a stem

leaflet

segment of a compound leaf, may be further divided or not; see also pinna

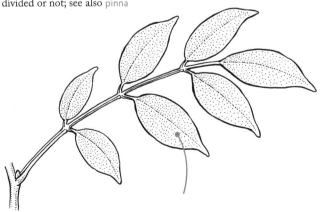

leaf scar

mark left on the stem from where the leaf
was attached, contains bundle scars

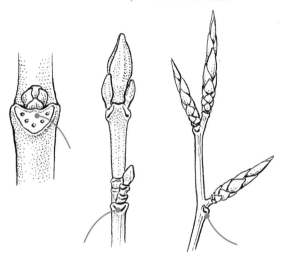

lemma

the lower/outer of two bracts subtending a
grass (Poaceae) floret, the other being the
palea

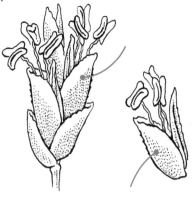

legume

1. dry, dehiscent unilocular fruit opening
along two lines of suture, derived from a
unicarpellate pistil; 2. vernacular name for
any plant in the bean family (Fabaceae); the
name comes from the old, and still accepted,
name for the family, Leguminosae

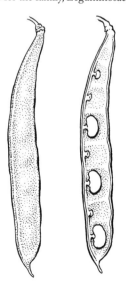

lenticel

raised linear to circular corky marking on
stems that allows for gas exchange

lenticular

shaped like a lentil, i.e., round and convex on
both sides

SYNONYM biconvex

lepidote
covered in small scales

liana
woody vine

ligneous, lignified
woody

ligulate
1.strap- or tongue-shaped; 2. having a ligule
SYNONYM 1. lingulate

ligulate flower
flower with an elongated fused corolla on one side, found in sunflower family (Asteraceae) inflorescences, together they form what look like the petals of these "false flower" inflorescences
SYNONYM ray flower
ANTONYM disk flower

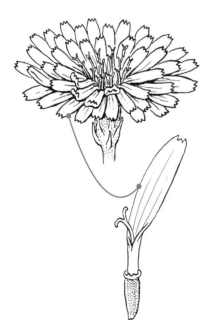

ligule
strap- or tongue-shaped structure; e.g., the projection at the top of the sheathing part of a grass (Poaceae) leaf by the base of the free blade, or the elongated fused corolla lobes in sunflower family (Asteraceae) ray flowers
SYNONYM ray

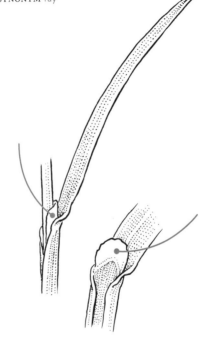

limb
expanded flat portion of a petal, leaf, or fused corolla

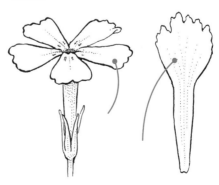

linear

narrow and long with parallel sides

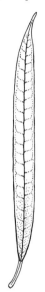

lobe

a rounded segment of something, as in a leaf or calyx

lingulate

strap- or tongue-shaped
SYNONYM ligulate

lip

1. central, usually lowest and largest petal of an orchid, may be cup-like; 2. lower prominent petal segment of a bilabiate flower, as in the mint family (Lamiaceae)
SYNONYM 1. labellum; 2. labium

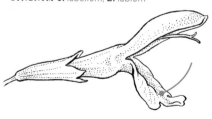

lithophyte

a plant that grows attached to a rock

lobed

having lobes, as in a leaf or stigma

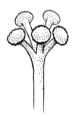

locule, loculus

chamber within an ovary, anther, sporangium, or fruit; in ovary and fruit, usually corresponding to a carpel

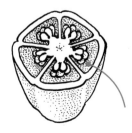

loculicidal

fruit opening to release seeds at the wall of the chamber; see also circumscissile, poricidal, septicidal

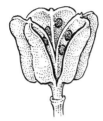

lodicules

two very small, usually flat structures subtending the flower inside of the lemma in grass (Poaceae) florets; presumably, these are the remnant perianth

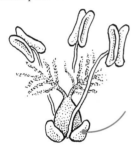

loment, lomentum

elongate fruit that is constricted and fused between the seeds and dehisces into corresponding one-seeded sections, derived from a unicarpellate pistil; fruit of some species in the bean family (Fabaceae)

long-day plant

plant that requires more than 12 hours of light per day to grow and reproduce

ANTONYM short-day plant

longitudinal section

cut along the main axis, abbreviated as l.s.

ANTONYM cross section

l.s.

longitudinal section

ANTONYM x.s., cross section

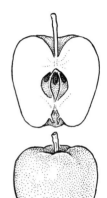

long shoot

stem with well-spaced nodes separated by long internodes, constitutes the majority of stems

ANTONYM brachyblast, short shoot, spur

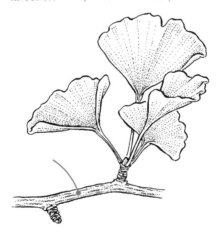

L

lyrate

pinnatifid with the terminal lobe rounded and much larger than subtending lobes

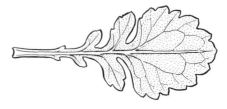

M

macro-

prefix meaning large

macrophyll

leaves with multiple branched veins, such as those of oaks (*Quercus*) and ginkgos (*Ginkgo*)

SYNONYM megaphyll

ANTONYM microphyll

maculate

marked with splotches or dots

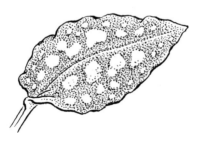

male flower

flower bearing fertile male structures (stamens) and no, or only infertile, female structures (pistils)

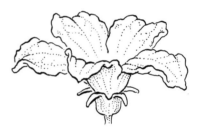

male sporophyll

male reproductive structure bearing pollen; e.g., the scales on male cones of pines (*Pinus*); see also stamen

marcescent

in reference to petals, sepals, and leaves, remaining attached although withered, as in the leaves of some beech trees (*Fagus*) and succulents

margin

edge, as of leaves, petals, and sepals

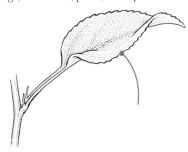

marginal

at, attached to, or near the edge

marginal placentation

ovules attached to the ovary wall in a simple pistil, such as in legumes (Fabaceae)

mast

edible fruits of forest trees, such as beechnuts (*Fagus*) and acorns (*Quercus*); most commonly used in the phrase "mast year" to indicate a year in which these fruits are produced in particularly large amounts

mat-forming

plant growth that results in a dense ground-covering

maturity

point at which an organ is fully developed, such as a ripe fruit or a fully open and functioning flower

medial, median

of or in the middle

medifixed

attached at the middle, as with filaments attached to the middle of anthers; see also basifixed, dorsifixed
SYNONYM versatile

mega-

prefix meaning large

megaphyll

leaves with multiple branched veins, such as those of oaks (*Quercus*) and ginkgos (*Ginkgo*)
SYNONYM macrophyll
ANTONYM microphyll

megasporangium

structure that bears female spores (megaspores)

megaspore

the larger, female spores found in
heterosporous plants, which include all seed
plants, certain aquatic ferns (e.g., *Azolla*,
Marselia, and *Salvinia*), and two lycophytes
(*Isoetes* and *Selaginella*)

membranous, membranaceous

very thin, nearly transparent

mericarp

individual section of a schizocarp, derived
from a single carpel in a syncarpous pistil;
e.g., members of the mallow (Malvaceae) and
carrot (Apiaceae) families

SYNONYM COCCUS

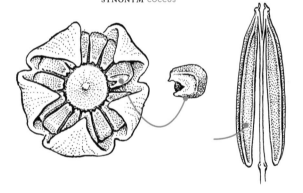

meristem

location of cell production that is responsible
for the overall growth in height, length,
and width of plants, located at the tips of
branches and roots as well as just inside the
bark

-merous

suffix meaning number of parts or sets,
usually applied to floral whorls

mesocarp

middle layer of the fruit wall (pericarp) ; e.g.,
the flesh of a peach (*Prunus persica*)

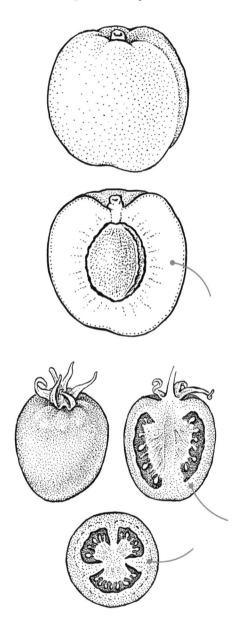

mesophyte

plant adapted to growing with average water availability; see also hydrophyte, xerophyte

microphyll

leaves with a single, usually unbranched vein, such as in horsetails (Equisetaceae) and spikemosses (Selaginellaceae)

ANTONYM macrophyll, megaphyll

microporangium

structure that bears male spores (microspores)

micropyle

opening between the integuments through which the pollen tube grows, visible on some seeds

microspore

the smaller, male spores found in heterosporous plants, which include all seed plants, certain aquatic ferns (e.g., *Azolla*, *Marselia*, and *Salvinia*), and two lycophytes (*Isoetes* and *Selaginella*)

midrib, midvein

central (primary) vein of leaves, usually more prominent than the lateral (secondary) veins

monadelphous

stamens fused together by their filaments into one mass (often a column)

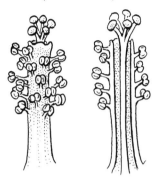

monandrous

having one stamen

mono-

prefix meaning one

monocarpic

flowering and fruiting only one time, then dying; e.g., century plant (*Agave americana*)

ANTONYM polycarpic

monochasium

cymose inflorescence with one single flower (simple) or multiple branching units (compound) produced on one side of a main axis

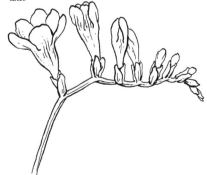

M

monocot

shortened name (from monocotyledon) for
the group of plants that usually have one seed
leaf (cotyledon), flower parts in multiples of
three, and parallel-veined leaves
ANTONYM dicot

monocotyledonous

having one seed leaf (cotyledon)

monoecious

having unisexual flowers of both sexes borne
on the same individual; illustrated in detail
here are smaller male (staminate) flowers and
a large female (pistillate) flower
ANTONYM dioecious

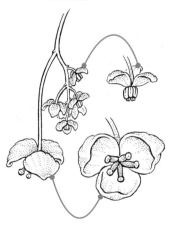

monopodial

having a single axis of growth with lateral
branching occurring on either side of the
main axis, most commonly used to describe
the growth of inflorescences but also to
describe vegetative growth, such as in some
orchids (Orchidaceae); see also sympodial

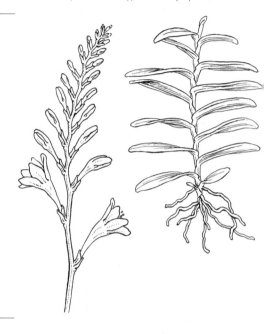

monotypic

having one type; e.g., a genus with only one
species or a family with only one genus

montane

living in the mountains

moss

non-vascular land plant in the division Bryophyta, often found in wet areas on the surface of soils, rocks, or tree trunks

motile

capable of movement; e.g., bryophyte and fern sperm

mottled

marked with splotches or dots of a different color

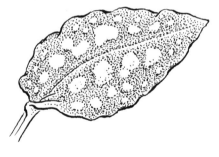

mouth

the opening of a tubular structure, such as a fused corolla

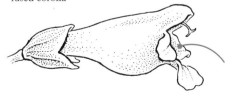

mucilage

thick, slimy, or gelatinous substance inside plants; e.g., the sap of *Aloe vera*

mucilaginous

slimy

mucro

a short, stiff, sharp point, as on a leaf apex or lobe; see also cusp

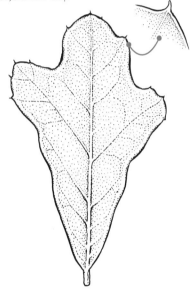

mucronate

coming to an abrupt, short, stiff, sharp point

SYNONYM cuspidate

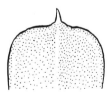

M

mucronulate

coming to an abrupt, very short, stiff, sharp point

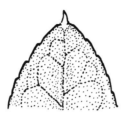

multi-

prefix meaning many

multicarpellate

having many carpels

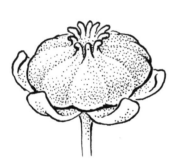

multilocular

having many locules

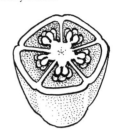

multiple infructescence

fruit derived from an entire inflorescence, may be fleshy or dry; e.g., sweetgum (*Liquidambar styraciflua*)

SYNONYM syncarp

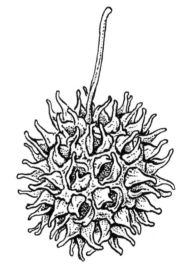

mutualism

a relationship in which two organisms live fused together or in very close proximity (symbiosis) and both benefit from the connection

mycoheterotroph

plant that derives nutrients from fungi; all plants previously thought of as saprophytic are actually mycotrophic and derive nutrients from green plant associates of the fungi on which they are parasites; e.g., Indian pipe (*Monotropa uniflora*)

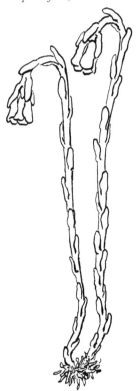

myrmecophyte

a plant that has a mutualistic relationship with ants

SYNONYM ant-plant

M

mycorrhiza

(plural mycorrhizae) fungus growing attached to a plant's roots and forming a symbiotic relationship with the plant; the fungus expands the plant's root system and delivers water and nutrients to the plant in exchange for sugars

N

naked

lacking a structure or structures normally present, such as a tree lacking leaves

nascent

beginning to develop, showing potential; as with introduced species with potential to become invasive

native

originating from a particular geographical or geological area
SYNONYM indigenous

naturalized, naturalised

established and reproducing non-native species; more established than adventive species

nectar

sticky, sugary fluid produced in various organs, such as flowers and leaves, as a reward/attractant, usually for pollinators but sometimes for insects, such as ants, that are associated with the plant

nectar guides

indicating markings (lines, spots, blotches) that direct pollinators to flower nectaries, may be invisible to human eyes except under ultraviolet (UV) light

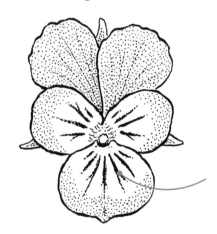

nectariferous

producing nectar

nectary

nectar-producing organ, gland, or tissue

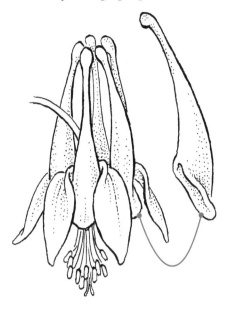

needle

long, very narrow leaf; leaves of many gymnosperms

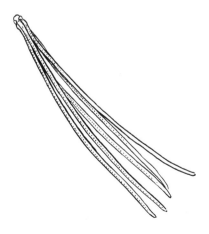

neotropics

the tropical areas of the Americas

nerve

vascular tissue in leaves or leaf-like structures such as bracts, petals, sepals, and stipules, may be branching or not

SYNONYM vein

net-veined, netted

having branched veins that connect to form an intricate pattern

SYNONYM reticulate

nitrogen

essential plant nutrient, abbreviated as N, the first number in fertilizer content

nitrogen fixation

process of conversion of atmospheric nitrogen into a form that can be absorbed by plants; often accomplished by bacteria (many of which were formerly known as blue-green algae) that live in a mutualistic relationship with plants such as legumes (Fabaceae)

nocturnal

occurring or active at night, as with the flowering of moonflower (*Ipomoea alba*) and many cacti

N

nodding

hanging or bent downward, applied most
often to flowers

SYNONYM cernuous

node

area of leaf attachment on the stem, may
have a leaf, leaf scar, or branch

ANTONYM internode

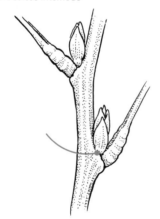

nodule

rounded knob, as with those on the roots
of many legumes (Fabaceae); see also root
nodule

nuciferous

producing nuts

numerous

greater than 10, often used to describe floral
parts; e.g., "stamens numerous"

nut

dry, indehiscent, unilocular fruit usually with
one seed and a hard pericarp; e.g., the acorns
of oaks (*Quercus*)

nutlet

1. small nut, as found in members of the
mint (Lamiaceae) and borage (Boraginaceae)
families; 2. another term for the achene fruit
of sedges (Cyperaceae)

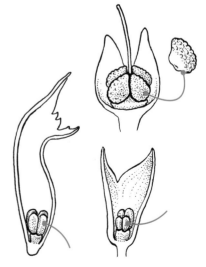

ob-
prefix meaning inverse

obconic, obconical
shaped like a cone, attached at the pointed
end

obcordate
1. heart-shaped with the widest point at
the top; 2. leaf apex with two rounded sides
separated by a broad indentation, like the top
of a heart

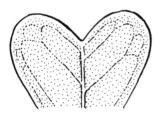

obdeltoid
equilateral triangle–shaped with one of the
flat sides on the top and attached at the
opposite point

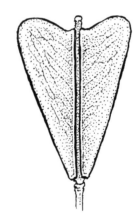

obdiplostemonous
having two distinct sets of stamens, the outer
whorl opposite the petals and the inner
whorl opposite the sepals
ANTONYM diplostemonous

oblanceolate

lance-shaped with the widest point toward
the top

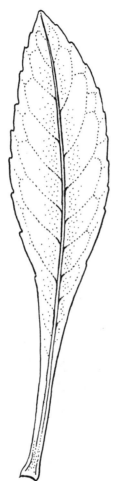

oblique

having two halves or sides that are unequal
in size and/or shape, usually applied to leaf
bases

SYNONYM asymmetrical

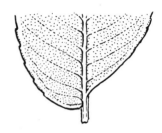

oblong

at least one and a half times as long as wide
with parallel sides

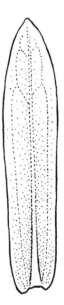

obligate

dependent on certain conditions, such as
the presence of another organism for some
parasites

obovate

egg-shaped with the broadest point at the top

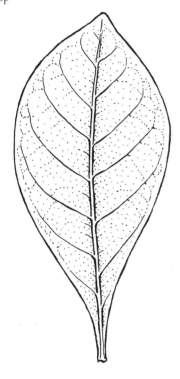

obtuse

rounded apex or base with curved sides forming a >90° but <180° angle

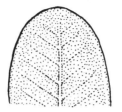

ocrea

(plural ocreae) stipules fused together into a sheath around the stem, as in many members of the buckwheat family (Polygonaceae)

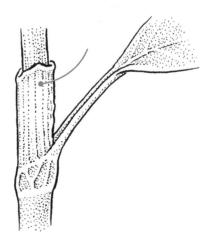

obovoid

three-dimensionally egg-shaped, broadest at the apex

obsolete

underdeveloped, reduced in size and not functional; e.g., non-functional, reduced stamen (staminode) in a flower

SYNONYM rudimentary, vestigial

O

octo-

prefix meaning eight

odd-pinnate

pinnately compound with an odd number
of leaflets, terminating in a single leaflet; see
also even-pinnate, paripinnate

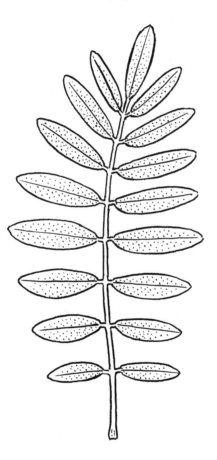

offset

shoot growing from the base of a main trunk
or stem, usually horizontal and useful in
propagation

offshoot

shoot growing off of a main trunk or stem

oligo-

prefix meaning few

open pollination

process of the free transfer of pollen from
one plant to another via insect, bird, wind,
water, or other natural mechanism; see also
chasmogamous

operculum

(plural opercula) small lid, as on a moss
capsule or eucalyptus flower

opposite

occurring across from each other, as with leaves paired two per node on a stem and stamens aligned with petals in a flower

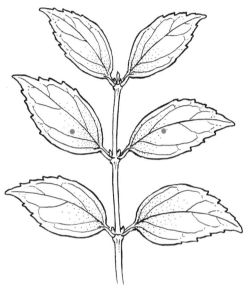

orbicular

round

SYNONYM circular

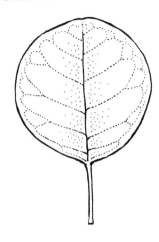

order

taxonomic rank above family and below class; plant order names end in "-ales"

organ

refers to external functional structures, such as roots, stems, leaves, flowers, and fruits

ornamental

plant cultivated for its appearance

ornithophilous

pollinated by birds

ortet

original parent plant from which propagules arose that grew additional, genetically identical (clonal) plants

ortho-

prefix meaning straight

outcross

selectively transfer pollen produced by one plant to the stigma of another

ovary

ovule-bearing section of a pistil, forms the fruit wall

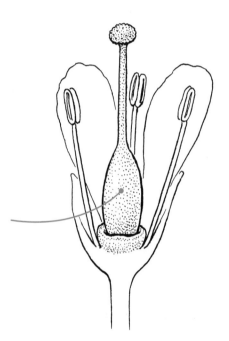

ovate

egg-shaped, broadest at the base

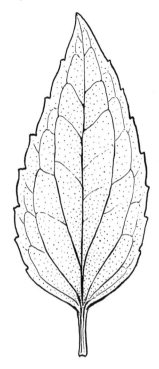

ovoid

three-dimensionally egg-shaped, broadest at the base

ovule

megasporangium, contains an egg cell and becomes the seed, located inside the ovary

P

pachy-
prefix meaning thick

pachycaul
thick-trunked with few to no branches,
usually applied to baobabs (*Adansonia*) and
other plants with bottle-shaped trunks

pad
segment of a cactus stem, may be used in
vegetative propagation

palate

projecting part on the lower lip of a bilabiate corolla; e.g., the bulging central portion that nearly closes the throat of a snapdragons (*Antirrhinum*)

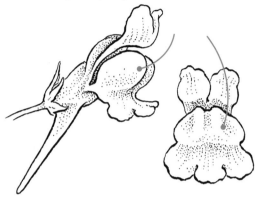

palmate

having veins, lobes, leaflets, or dissections all arising from a single point, usually at the top of the petiole, like fingers from a palm
SYNONYM digitate

palea

the upper/inner of two bracts subtending a floret in the grass family (Poaceae), the other being the lemma

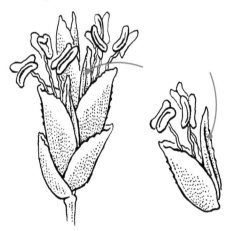

palmately compound

compound leaf with dissections all arising from a single point, usually at the top of the petiole, like fingers from a palm

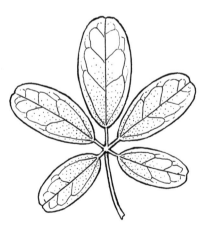

paleotropics

the tropical areas of Africa, Asia, and the Pacific, excluding Australia and New Zealand

palmately lobed, palmatifid

having lobes all arising from a single area on the blade, like fingers from a palm

palmatisect

very deeply palmately lobed

pandurate

shaped like a fiddle, i.e., with rounded ends and a contracted center

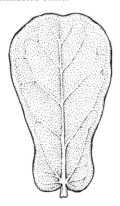

panicle

inflorescence with pedicellate flowers borne on branches arising from an elongated central axis

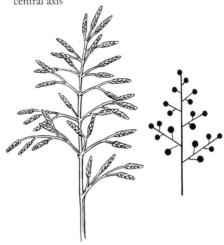

pantropical

occurring in all tropical regions of the world

papilionaceous

butterfly-shaped, usually in reference to flowers characterized by having a large upper banner petal, two lateral wing petals, and two petals fused to form a single lower central keel; typical of papilionoid legumes (Fabaceae subfamily Papilionoideae)

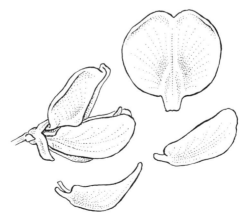

P

papilla

(plural papillae) short, rounded, nipple-like protuberance

papillate

having papillae

papillose

having small papillae

pappus

1. modified calyx of the sunflower family (Asteraceae) that may consist of very short to long modified sepals in the form of bristles, awns, or scales, and that sometimes facilitates wind dispersal; 2. dense cluster of hairs attached to the end of a milkweed (*Asclepias*) seed that facilitates wind dispersal, more correctly called a coma

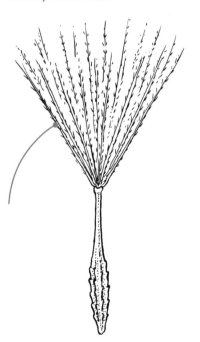

parallel-veined

having veins that run more or less side by side the length of the leaf

parasite

organism that is attached to and draws water and/or nutrients from another organism (the host), may be partially or fully dependent on the host; e.g., Indian pipe (*Monotropa uniflora*)

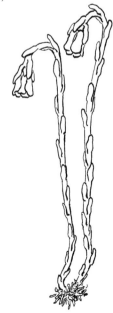

parastichy

spiral formed by tracing the points of
attachment of organs on an axis, such as
leaves along a stem or scales along the central
axis of a cone

parietal placentation

ovules attached to the ovary wall in a usually
unilocular compound pistil (multicarpellate
ovary)

paripinnate

pinnately compound with an even number of
leaflets, terminating in a pair of leaflets; see
also imparipinnate, odd-pinnate

SYNONYM even-pinnate

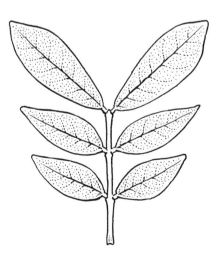

parthenocarpy

fruit production without fertilization or seed
development

parthenogenesis

seed production without fertilization

patent

spreading outward, as with petals from the
floral axis or lower branches from a tree trunk

pectinate

resembling a comb in having narrow, closely
spaced segments

pedate

palmately divided leaf with the lower leaflets
further split in two

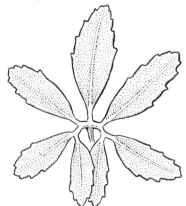

pedately lobed
deeply palmately lobed leaf with the lower lobes further split in two

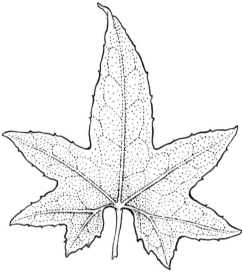

pedicellate
having a stalk, applied to flowers

peduncle
the stalk of a solitary flower or an entire inflorescence

pedicel
the stalk of an individual flower within an inflorescence

pellucid

translucent or transparent; e.g., gland dots in leaves and fruit exocarp of lemons and oranges (*Citrus*)

peltate

having a centrally attached stalk or petiole like an umbrella; e.g., the leaves of water lotuses (*Nelumbo*)

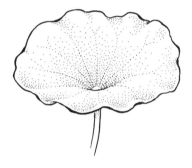

pendent, pendulous

hanging or bending downward

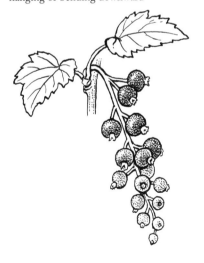

penta-

prefix meaning five

pepo

unilocular berry with hard exocarp, characteristic of the gourd family (Cucurbitaceae)

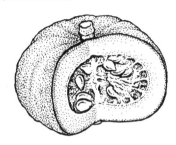

perennial

living and reproducing for more than two years, i.e., sets seed multiple times over its life

SYNONYM polycarpic

perfect

flowers that have functioning female and male reproductive parts

perfoliate

leaf, stipule, or bract base fused, so as to appear to be pierced by the stem

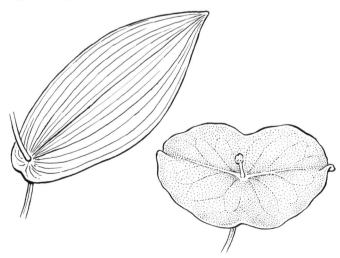

P

peri-

prefix meaning about or surrounding

perianth

collective term for calyx (sepals) and corolla
(petals)

SYNONYM floral envelope

pericarp

fruit wall, derived from the ovary wall and
consisting of up to three layers: exocarp,
mesocarp, and endocarp

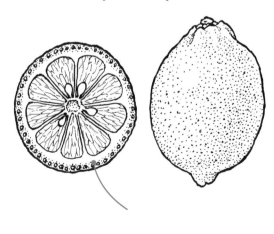

perigynium

(plural perigynia) covering surrounding the
pistil in sedges (*Carex*), often hard

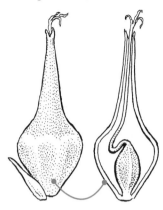

perigynous

flower with a superior ovary and a floral cup
(hypanthium)

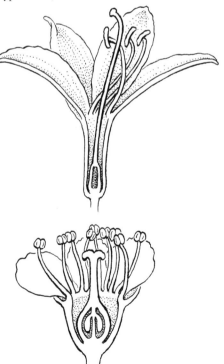

persistent

remaining attached beyond what is normal
for that type of structure, as with the calyx in
strawberries (*Fragaria*) or rose hips (*Rosa*)

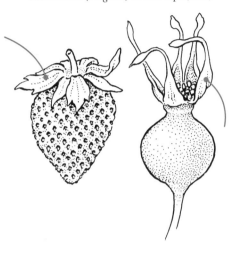

petal

individual component of the second whorl
of the flower (the corolla), which is usually
colorful and functions to attract pollinators
and facilitate pollination; variously shaped;
illustrated in detail below (clockwise from
top) are obcordate, spatulate, bifid, and
apiculate petals

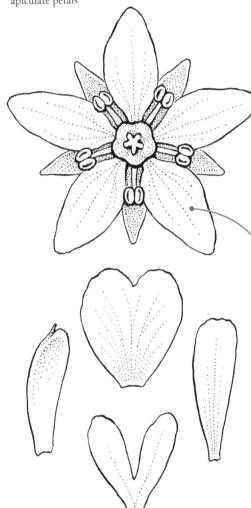

personate

with two lips (as of a bilabiate corolla) that
stay nearly fully closed, requiring pollinators
to push their way in

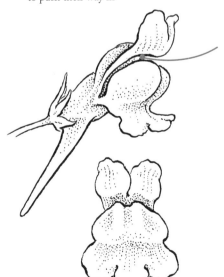

petaloid

petal-like, may apply to any floral whorl, as in petaloid sepals (top) or petaloid stamens (below)

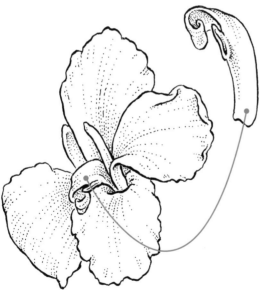

petiolate

having a petiole

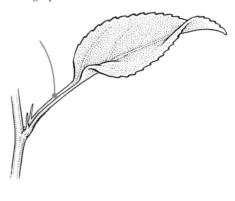

petiole

leaf stalk that connects the blade to the stem

petiolule

leaflet stalk in a compound leaf

phanerogam

plant reproducing with seeds, not spores

ANTONYM cryptogam

-phore

suffix meaning stalk

phosphate

essential plant nutrient, abbreviated as P, the second number in fertilizer content

photoperiodism

growing or flowering in response to the length of light and/or dark periods

photosynthesis

process through which light energy from the sun is converted to chemical energy stored in sugars, occurs in chloroplasts

phototropism

growth or orientation of shoots, flowers, and leaves toward light

phyllary

one of the many bracts of the involucre subtending the capitulum inflorescence in the sunflower family (Asteraceae)

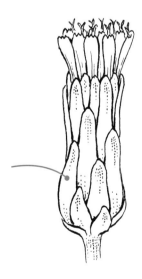

phylloclade

stem that looks and functions like a leaf

SYNONYM cladode, cladophyll

phyllode

leaf-like structure formed from a laterally expanded petiole without a blade or with a very reduced blade, as in some mimosas and wattles (*Acacia*); also applied to the flattened, non-pitcher leaves of pitcherplants (*Sarracenia*)

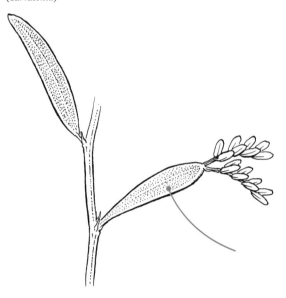

phyllotaxy

arrangement of leaves along the stem

pilose

covered in soft, long hairs

pinna

(plural pinnae) primary segment of a pinnately compound leaf; may be further divided or not; see also leaflet

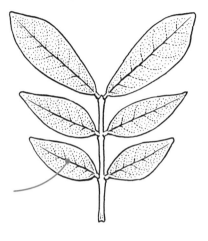

pinnatifid

pinnately lobed blade whose sinuses extend halfway or slightly more to the midrib

pinnate

leaf with veins, lobes, leaflets, or dissections arising along a central elongate axis

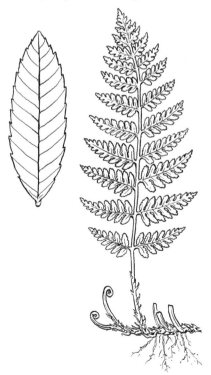

pinnatisect

pinnately lobed blade whose sinuses extend nearly to the midrib

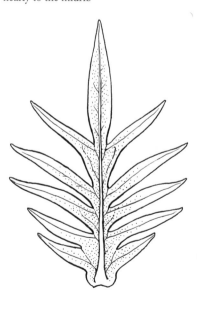

pinnule

ultimate segment of a leaf that is more than once-pinnately compound, as in some fern fronds

pit

the hard middle part of a fleshy fruit, may be an endocarp, as in a peach (*Prunus persica*), or just a hard seed, as in an avocado (*Persea americana*)

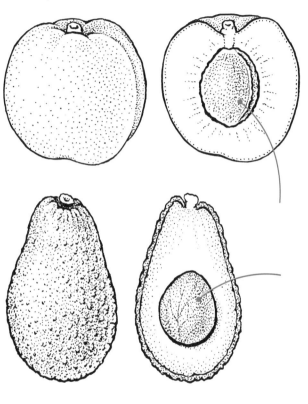

pistil

individual component of the innermost whorl of the flower (the gynoecium), each is made up of one to many fused carpels and normally consists of stigma(s), style(s), and ovary

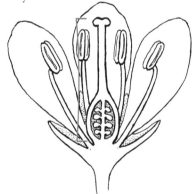

pistillate

having female reproductive structures (pistils) and lacking male structures (stamens)

pith

soft, spongy tissue in the very center of the stems and monocot roots of vascular plants

P

placenta

tissue to which ovules are attached inside an ovary and seeds are attached inside a fruit

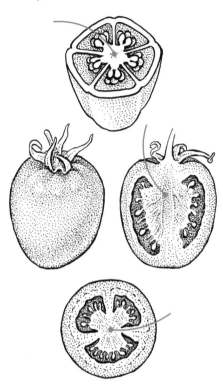

placentation

arrangement of placentas in an ovary; placenta location is most easily determined by locating the point(s) of attachment of the ovules or seeds

plane

flat surface

plano-convex

flat on one side and rounded outward on the other

plantlet

small plant, used in reference to those formed vegetatively on another plant naturally or from another plant through propagation

pleated

folded like a fan in regular, longitudinal pleats; e.g., the corolla of morning glories (*Ipomoea*), the fronds of palmettos (*Sabal*)

SYNONYM plicate

pleio-

prefix meaning more

pleiochasium

compound cyme with more than two branches produced at the first juncture of the main axis

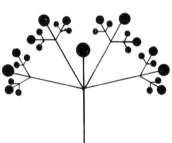

pleiomerous

having more than the usual number of parts in a floral whorl; e.g., roses (*Rosa*) that have a proliferation of petals

SYNONYM doubled

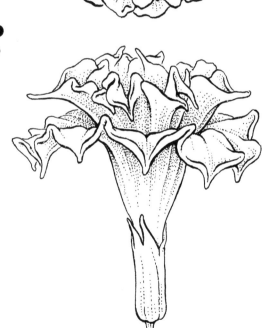

plicate

folded like a fan, a form of aestivation in which young flower parts are folded in regular, longitudinal pleats; e.g., morning glories (*Ipomoea*)

SYNONYM pleated

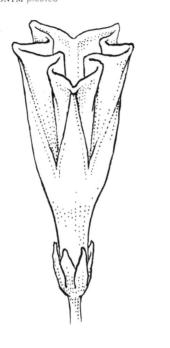

plumule

first shoot in a germinating seed

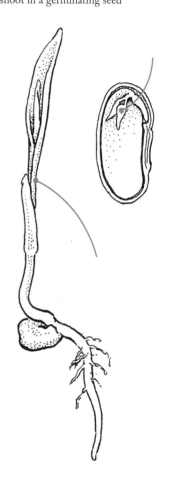

plumose

feather-like in structure and appearance

pneumatophores

vertical roots that are the site of gas exchange for otherwise inundated root systems, common in mangroves (*Avicennia*, *Rhizophora*)

pod

unspecific name for dry, dehiscent fruits such as legumes, capsules, silicles, siliques, and follicles

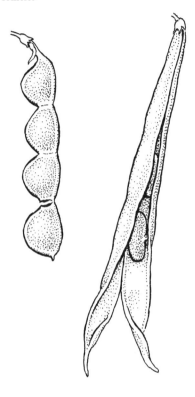

pollard

style of tree pruning in which tips of trunk and/or branches are cut back, encouraging new growth

pollen

microspore (when young), male gametophyte (when mature) of seed plants, carries sperm cells to ovules for fertilization; produced in anthers in flowering plants (angiosperms) and on scales of male cones in gymnosperms

pollinarium

in milkweeds (*Asclepias*), two pollinia, two translator arms, and a sticky corpusculum that attaches to a pollinator for transport

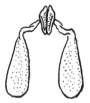

pollination

process by which pollen is delivered to the ovule of a gymnosperm or the stigma of an angiosperm

pollinator

organism (e.g., bird or insect) or other agent (e.g., wind or water) that transfers pollen from one plant to another; most commonly applied to organisms

pollinium

(plural pollinia) all the pollen in a single anther sac held together in a single mass for transfer by a pollinator, as found in milkweeds (*Asclepias*) and orchids (Orchidaceae); in milkweeds, part of a pollinarium

poly-

prefix meaning many

polycarpic

living and reproducing for more than two years, i.e., sets seed multiple times over its life

SYNONYM perennial

ANTONYM monocarpic

P

polychasium

compound cyme with more than two
branches produced by each axis

polygamous

having male, female, and bisexual flowers on
the same individual plant

polymorphic

having many forms, applied to whole
organisms or individual structures

polyploid

having more than two sets of chromosomes
(e.g., 3n, 4n, 5n, 6n); see also diploid, haploid,
tetraploid

pome

false fruit consisting of an expanded, fleshy
hypanthium fused with the pericarp (core);
called false because the majority of the tissue
is derived from the hypanthium rather than
the ovary; e.g., apples (*Malus*) and pears
(*Pyrus*)

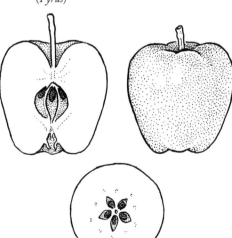

pore

small opening, as in anthers (e.g., Ericaceae)
or capsules (e.g., poppies, *Papaver*)

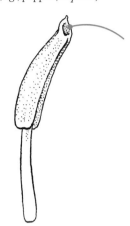

poricidal

opening through one or more pores, as
with some anthers or capsules; see also
circumscissile, loculicidal, septicidal

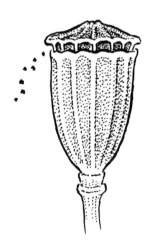

potash

name for the form of potassium used in fertilizer

potassium

essential plant nutrient, abbreviated as K, the third number in fertilizer content

prickles

sharp, pointed epidermal projections

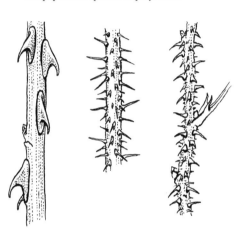

primary

first segment or branch; e.g., largest vein(s) in a leaf, leaflets in a pinnately compound leaf

primocane

a fruiting shrub's first-year stems, which may produce fruit late in the first growing season or wait until the next year to produce fruit; in some plants, such as raspberries and blackberries (*Rubus*), individuals with primocanes are called fall-bearing
ANTONYM floricane

procumbent

growing along the ground without rooting at the nodes

proliferous

vegetatively reproducing by buds or plantlets usually produced on a leaf or flower

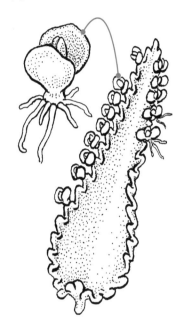

prominent

conspicuous, as with a midvein on a leaf

propagation

the process of growing a plant from spore, seed, or vegetative stock from a parent

P

propagule

product or part of a plant capable of growing
a new plant; e.g., bud, spore, seed, cutting

prop root

adventitious root emerging from the lower
part of a trunk and acting as a structural
support for the tree

SYNONYM anchor root, brace root, stilt root

prostrate

lying or growing flat along the ground

SYNONYM recumbent

protandrous

anther mature and releasing pollen before
the stigma is mature and receptive to pollen,
helps prevent self-pollination

protogyny

stigma mature and receptive to pollen before
the anther is mature and releasing pollen,
helps prevent self-pollination

proximal

base, the end closest to the point of
attachment

ANTONYM distal

pseudanthium

inflorescence that closely resembles a single
flower; e.g., dogwood (*Cornus florida*),
Euphorbia cyathia, Asteraceae heads/capitula

SYNONYM false flower

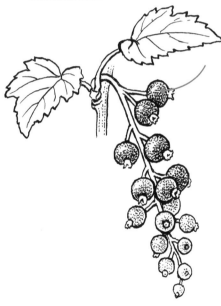

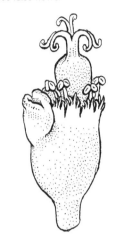

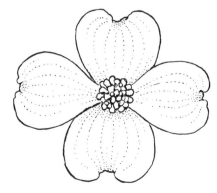

P

pseudo-

prefix meaning false or resembling

pseudobulb

swollen stem resembling a bulb, as in some orchids (Orchidaceae)

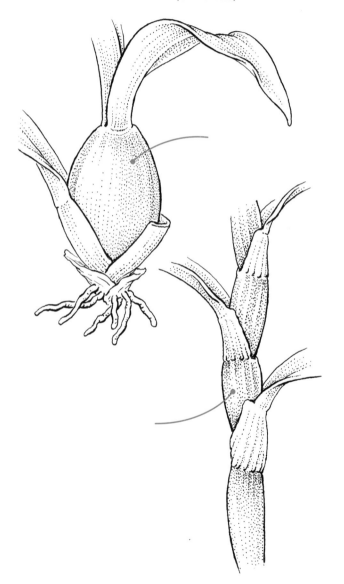

pseudocarp

seed-bearing structure resembling and often mistaken for a fruit but for which the majority of the tissue is non-ovary (may be from such structures as a hypanthium or receptacle); e.g., rose hips (*Rosa*)

SYNONYM anthocarp, false fruit

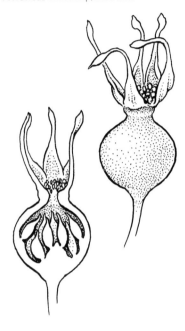

pseudocopulation

pollination strategy of some orchids in which floral parts mimic a female insect, tricking the male insect into attempting to mate with the flower

SYNONYM sexual deception

pseudoterminal

axillary growth that appears terminal in origin, as with some stems for which extending growth originates from lateral buds

ptyxis
arrangement of leaves in bud; see also
aestivation

SYNONYM vernation

puberulent
having very small hairs

pubescence
hairiness

pubescent
having hairs

pulvinus
(plural pulvini) swollen base of petiole or
petiolule (in latter case, sometimes called
pulvinulous)

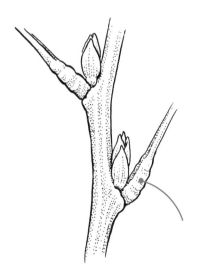

pup
a small plant produced vegetatively by
another plant

pyramidal
shaped like a pyramid

pyriform
shaped like a pear

P

punctate
spotted with pits or dots

quad-
prefix meaning four

quinque-
prefix meaning five

R

raceme

inflorescence with pedicellate flowers borne on an unbranched elongated central axis

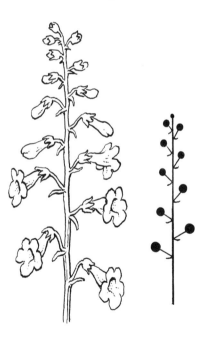

rachilla

a small or secondary rachis, usually applied to the axis of a spikelet of grasses or sedges

rachis, rhachis

the central axis of a branched or dissected organ, such as a pinnate leaf or an inflorescence

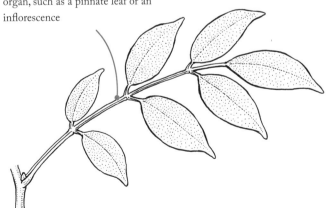

radially symmetrical

having multiple planes of symmetry such that any line drawn through the middle produces two mirror-image halves, usually applied to flowers

SYNONYM actinomorphic, regular

ANTONYM bilaterally symmetrical, irregular, zygomorphic

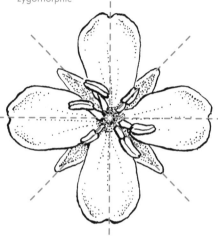

radicle

first root of a germinating seed

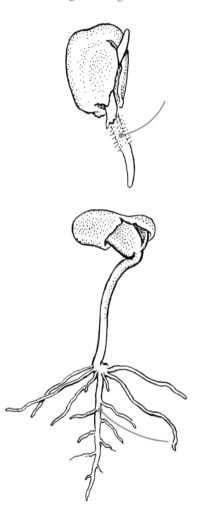

radiate

1. spreading outward, as with the stigmas of poppies (*Papaver*) or petals from the flower's center; 2. in the sunflower family (Asteraceae), head/capitulum inflorescence having ray/ligulate flowers

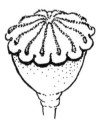

ramet

a vegetatively reproduced individual in a genetically identical colony called a genet

ramicaul

a single-leafed stem, as in *Pleurothallis* orchids

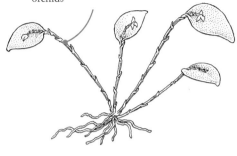

range

native geographic distribution

ratoon

stem that sprouts from the root of a perennial plant that has been cut down, as with sugarcane (*Saccharum*) after it is harvested, or trees after they are felled

ray

1. the elongated fused corolla of some flowers in the sunflower family (Asteraceae); 2. in wood, a band of tissue that runs perpendicular to the vascular tissue, creating cross markings across the vascular rings

SYNONYM 1. ligule

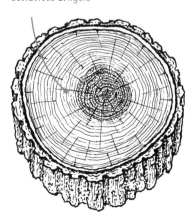

ray flower

flower with an elongated fused corolla on one side, found in sunflower family (Asteraceae) inflorescences, together they form what look like the petals of these "false flower" inflorescences

SYNONYM ligulate flower

ANTONYM disk flower

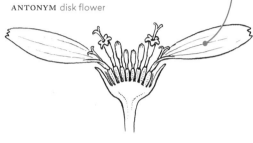

receptacle

1. in flowers and fruits, tissue to which all the floral whorls are attached; 2. in the sunflower family (Asteraceae), location on inflorescence where all the florets are attached

SYNONYM 1. torus

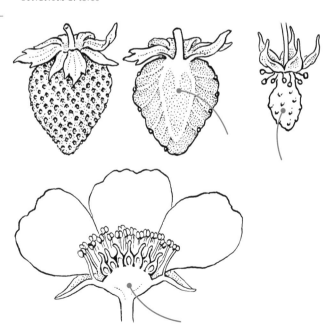

R

receptive
mature and ready to receive pollen grains, applied to the stigmas of pistils

recumbent
lying or growing flat along the ground
SYNONYM prostrate

recurved
curved backward toward the point of attachment

reflexed
bent backward toward the point of attachment

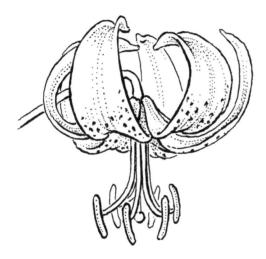

refoliate
grow leaves back after an unexpected loss from such things as disease, herbivory, or spring frost

reduplicate
folded from the base to the apex with the under (abaxial) surface facing itself
ANTONYM conduplicate

regular

having multiple planes of symmetry such that
any line drawn through the middle produces
two mirror-image halves, usually applied to
flowers

SYNONYM actinomorphic, radially symmetrical

ANTONYM bilaterally symmetrical, irregular,
zygomorphic

reniform

shaped like a kidney

repand

margin or surface shallowly wavy, more often
applied to surfaces

SYNONYM undulate

replum

1. persistent septum to which seeds are
attached in the middle of silicle and silique
fruits in the mustard family (Brassicaceae);
2. marginal placenta that separates from
the fruit wall at dehiscence, e.g., in *Mimosa*
(Fabaceae)

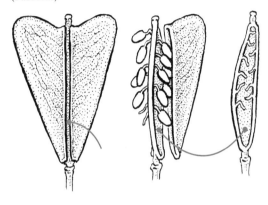

reseed

plant seeds again, usually used in reference to
grasses in lawns

resin

sticky exudate of some woody plants that is
insoluble in water

resupinate

pedicel twisted 180°, turning the flower
upside-down

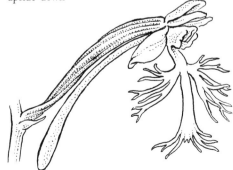

R

reticulate

having branched veins that connect to form
an intricate pattern

SYNONYM net-veined, netted

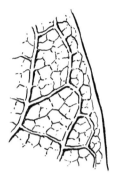

retrorse

pointing down or toward the base

ANTONYM antrorse

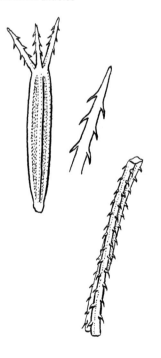

retuse

having a rounded apex with an abrupt and
shallow indentation in the center

SYNONYM emarginate

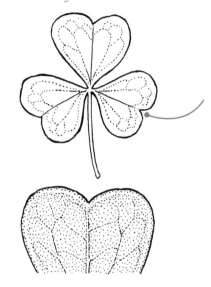

revolute

rolled downward toward the lower (abaxial)
surface

ANTONYM inrolled, involute

rhachis, rachis

the central axis of a branched or dissected organ, such as a pinnate leaf or an inflorescence

rhizome

an underground, usually horizontal stem, such as that of ginger (*Zingiber officinale*)

rhomboid, rhombic

shaped like a diamond

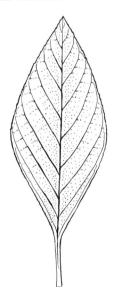

R

rhizomatous

having rhizomes

rib

conspicuous leaf vein, most commonly applied to primary veins

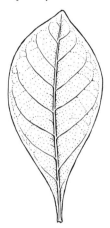

riparian

growing on the banks of rivers or streams

ripe

fully mature, as with fruit

root

the usually below-ground part of a plant that lacks leaves, stems, and nodes, organ through which most nutrients and water are absorbed

root ball

the roots and soil that remain attached to a woody plant dug up for transfer to a new location, often wrapped in burlap

rootbound

said of a plant, potted or otherwise confined, whose roots have filled the available space and grown in unusual and often unhealthy formations as a result

root crown

location on a plant where the stem and root systems meet

root nodule

rounded knobs on the roots of many legumes (Fabaceae) which house nitrogen-fixing bacteria

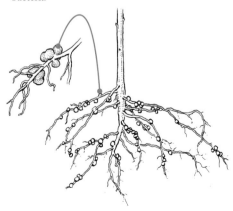

rootstock
plant onto which another plant (the scion)
is grafted

rosette
dense ring of leaves, or other organs, around
the base of a plant at or near the ground

rostellum
projection in an orchid flower's column
between the stigma and the anther, helps
prevent self-pollination

rotate
shaped like a disk, applied to a corolla
in which petals lack a floral tube and are
extended, forming a flat circular plane

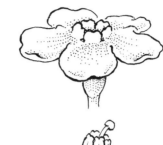

rudimentary
underdeveloped, reduced in size and not
functional; e.g., non-functional, reduced
stamen (staminode) in a flower
SYNONYM obsolete, vestigial

rufous, rufus
reddish brown, rust- or chestnut-colored
SYNONYM castaneous, ferruginous

rugose
wrinkled

ruminate
coarsely wrinkled, appearing as if chewed

R

runcinate

pinnatifid with the lobes pointed and
sloping toward the petiole; e.g., the leaves of
dandelion (*Taraxacum officinale*)

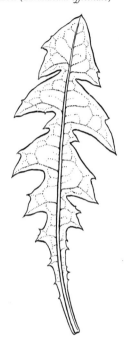

runner

horizontal, above-ground, creeping stem with
roots and shoots forming at nodes and the
tip, as with strawberries (*Fragaria*)

SYNONYM stolon

saccate

bag-shaped or consisting of bag-like
structures

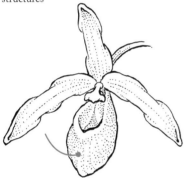

sagittate

arrowhead-shaped with basal lobes pointing
downward

salverform

trumpet-shaped with a long, narrow neck, as
with some fused corollas

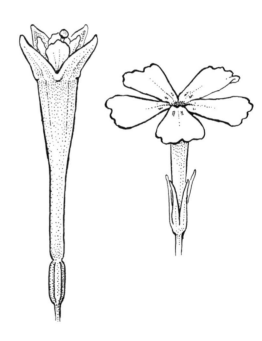

samara

dry, indehiscent fruit with wings formed
from an expansion of the pericarp; e.g.,
the fruit of ash (*Fraxinus*) and tulip trees
(*Liriodendron*)

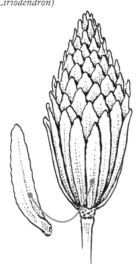

sap

the liquid within the vascular tissue of plants

sapling

young tree that has woody growth but which
is still quite flexible

saprophyte

fungus that derives nutrients from decaying
organic matter; all plants previously thought
of as saprophytic actually have a parasitic
relationship with a fungus and a green plant;
see mycoheterotroph

sapwood

outer, younger, lighter wood of trees; lies
outside of the heartwood and is usually
considered less desirable for woodworking

samaroid

samara-like

samaroid schizocarp

fruit derived from a two-carpellate ovary that
splits at maturity into two winged sections
(mericarps that resemble samaras); e.g., the
fruit of maples (*Acer*)

SYNONYM double samara

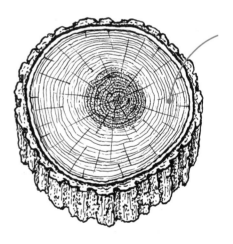

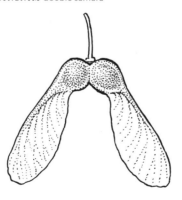

sarmentose

having long, thin runners (stolons)

SYNONYM flagellate

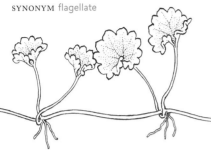

scabrous

rough, like sandpaper, to the touch

scale

1. type of epidermal projection (trichome) that is usually flattened and broad; 2. small leaf or leaf-like structure; 3. one of the fleshy or dried leaves of a bulb; 4. cone/strobilis segment, usually bearing spores, pollen, ovules, or seeds; 5. sap-sucking insect pest in the superfamily Coccoidea

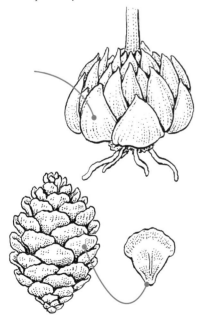

scandent

climbing or leaning so as to be dependent on other plants or structures for vertical growth support

scape

leafless flower or inflorescence stalk (peduncle) originating from a root, bulb, or corm, typically in plants that have a basal rosette of leaves; e.g., tulips (*Tulipa*), Japanese primrose (*Primula japonica*)

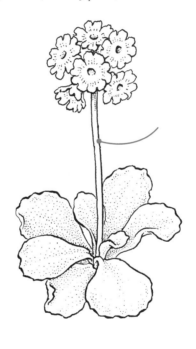

scapiform

closely resembling a scape but with leaves along the stalk

S

scar

1. mark left on the stem where the leaf was attached; 2. mark within the leaf scar on the stem from where the leaf's vascular tissue was attached; 3. mark left on the seed from where the funicle attached it to the ovary wall; 4. mark left on any organ from where damage was done to epidermal tissue

schizocarp

fruit derived from a two- to multicarpellate pistil that splits at maturity into individual segments (mericarps, equivalent to the carpels)

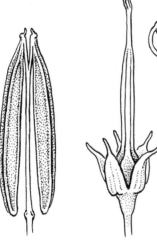
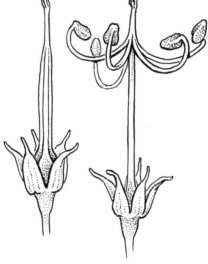

scion

plant cutting that is grafted onto a rootstock in order to grow a plant that has the above-ground characteristics of the cutting

scorpioid cyme

sympodial inflorescence with flowers borne alternating from side to side on the axis and resembling a scorpion's tail; can be difficult to distinguish from a helicoid cyme

scurf

a covering of scales on some plant organs, may cause dark or rough spots

secund

with parts on one side only; e.g., flowers on only one side of an inflorescence

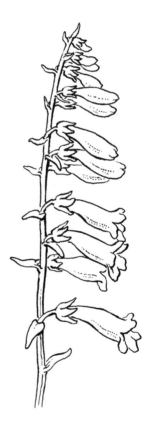

seed

sexual reproductive structure in which the
embryo is housed; a ripened ovule; e.g., the
pit of an avocado (*Persea americana*); the seed
of a sunflower (*Helianthus annuus*)

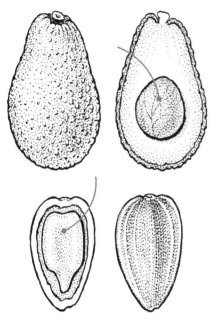

seed leaf

one of a seed's first leaves

SYNONYM cotyledon

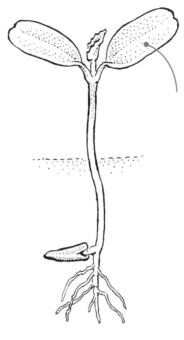

seed coat

layer of tissue covering a seed, derived from
the integuments surrounding the ovule

SYNONYM testa

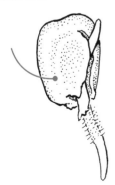

seedling

very young plant growing from a recently
germinated seed

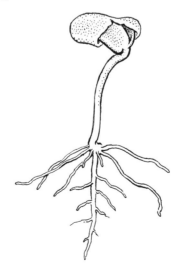

S

self-pollination, selfing
transfer of pollen from the plant's own anther
to a stigma on the same plant

sepal
individual component of the outermost
whorl of the flower (the calyx), leaf-like or
petal-like

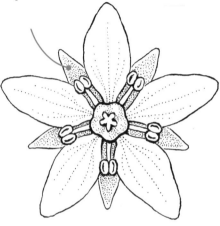

sepaloid
sepal-like in appearance

septicidal
opening at the septa to dehisce; see also
circumscissile, loculicidal, poricidal

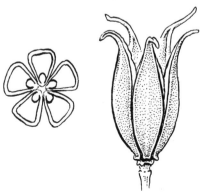

septum
(plural septa) wall between chambers
(locules) in an ovary or fruit

serotinous
having seed release triggered by an
environmental condition such as heat from
fire

serrate

having teeth pointing up toward the apex

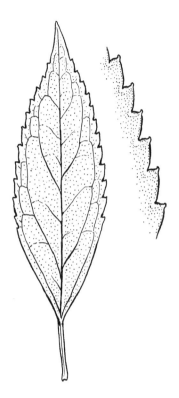

serrulate

having minute teeth pointing up toward the apex

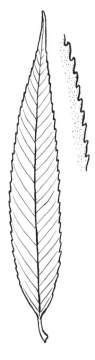

serration

individual tooth of a serrate margin, or the whole margin itself

sessile

lacking a stalk; e.g., leaf without a petiole

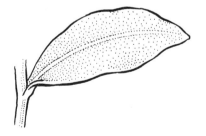

set

in plant propagation, a young transplant

setose

bearing bristles; bristly

sexual deception

pollination strategy of some orchids in which floral parts mimic a female insect, tricking the male insect into attempting to mate with the flower

SYNONYM pseudocopulation

sheath

usually flattened and elongated part of a structure that fully to partially covers another structure, as with the leaf base on the stem of some monocots

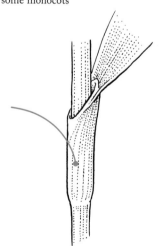

sheathing

forming a sheath, as with leaves sheathing a stem

shoot

stem, often applied to new stem growth

short-day plant

plant that requires more than 12 hours of darkness per day to grow and reproduce

ANTONYM long-day plant

short shoot

stem with highly compressed internodes that usually bears the leaves and reproductive structure; e.g., ginkgos (*Ginkgo*), apples (*Malus*)

SYNONYM brachyblast, spur

ANTONYM long shoot

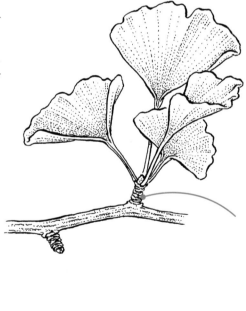

shrub

woody plant with multiple main stems, usually shorter than a tree
SYNONYM bush

shrublet

small shrub

sigmoid

shaped like the letter S

silicle

dry, dehiscent, short, wide, flattened fruit with two sides, the locules of which open to reveal a central persistent septum (replum); found in members of the mustard family (Brassicaceae)

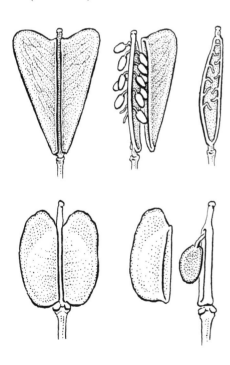

silique, siliqua

dry, dehiscent, long, narrow fruit with two sides, the locules of which open to reveal a central persistent septum (replum); found in members of the mustard family (Brassicaceae)

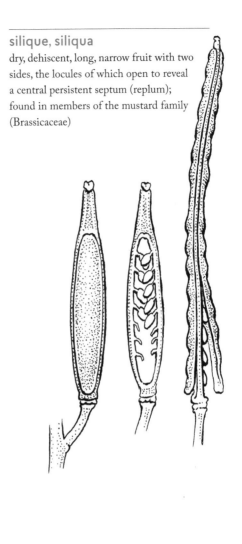

S

silk

the long, fine, soft styles of corn (*Zea mays*)
inflorescences and cobs

simple

1. an undissected leaf; 2. an unbranched
inflorescence

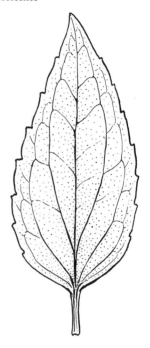

sinker

shoot growing downward from a bulb or
corm, ending in a new bulb or corm

SYNONYM dropper

sinuate, sinuous

having a wavy margin

sinus

the portion of a margin that dips in between two lobes or crenations

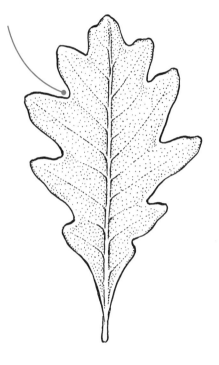

smooth

1. in relation to surfaces, even, not bumpy or rough; 2. in relation to margins, entire, not lobed or toothed

snag

1. a dead tree; 2. remaining portion of a branch that has otherwise been removed

solitary

singular, only one

sorus

(plural sori) clump of fern sporangia, usually freely borne on the underside of a fern leaf, or concealed by flaps or sections of tissue known as indusia and false indusia

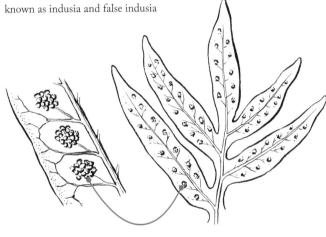

sp.

abbreviation for one species (singular)

spadix

unbranched inflorescence with flowers slightly sunken into an elongated, thickened axis; the inflorescence of aroids (Araceae)

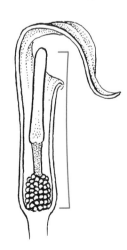

S

spathe

bract that subtends and/or partially
surrounds the spadix inflorescence of aroids
(Araceae)

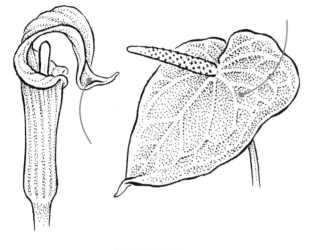

spatulate

shaped like a spatula

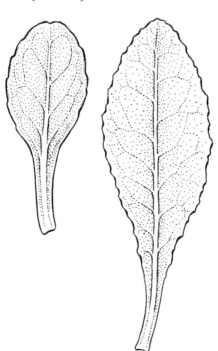

species

taxonomic rank below genus and inclusive of
any subspecies, varieties, or forms

spherical

round in three dimensions

SYNONYM globose, globular

spicate

having spikes or being in a spike

spike

inflorescence with sessile flowers borne on an
unbranched elongated central axis

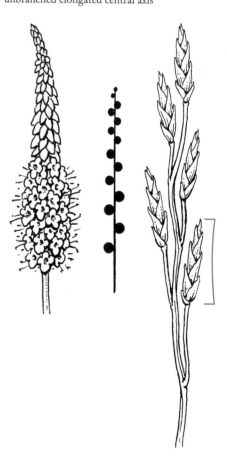

spikelet
small spike

spine
sharp, pointed modified leaf, leaflet, bract, sepal, or stipule

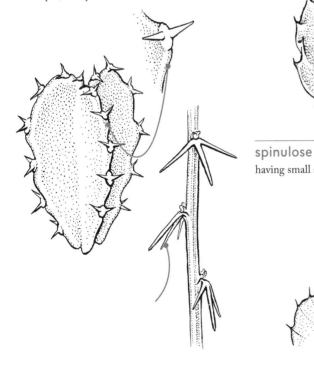

spinose, spiny
having spines

spinose tooth
tooth at the margin of a leaf or leaf-like structure that is pointy and sharp and resembles a spine

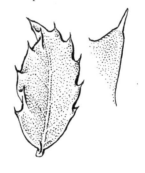

spinulose
having small spines

S

sporangium

(plural sporangia) spore-containing pouch or sac

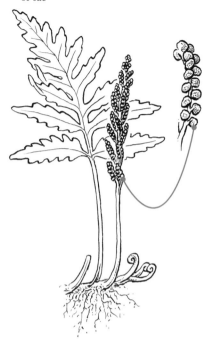

spore

reproductive unit and the first cell of the gametophytic stage of a plant's life cycle, usually unicellular and microscopic

sporophyll

sporangia-bearing specialized leaf; e.g., cone scale, pistil, stamen

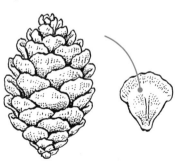

sporophyte

life cycle generation in which a plant has two sets of chromosomes (i.e., is diploid, 2n) and produces spores; the sporophyte is dominant in both time and size for vascular plants, making it their most conspicuous generation; e.g., tree, fern

ANTONYM gametophyte

sport

a shoot whose morphology does not match that of the rest of the plant; mutant shoot

spp.

abbreviation for more than one species (plural)

spring ephemeral

plant that grows, flowers, fruits, and dies back completely by mid summer

sprout

1. a seedling; 2. to send out new growth

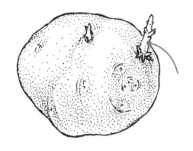

spur

1. hollow appendage on a flower, often containing nectar and a projection from or a modification of the perianth; 2. stem with highly compressed internodes that usually bears the leaves and reproductive structure, e.g., ginkgos (*Ginkgo*), apples (*Malus*)
SYNONYM 1. calcar; 2. brachyplast, short shoot
ANTONYM 2. long shoot

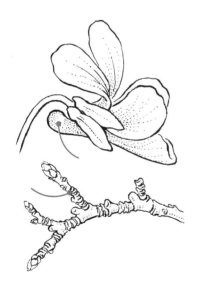

spurred

having a spur
SYNONYM calcarate

squam-

prefix meaning scales

squamose, squamate

covered in scales

stalk

structure subtending an organ such as a flower (called a pedicel or peduncle) or leaf (called a petiole), most often more narrow than the organ itself

stamen

male reproductive structure consisting of a filament (stalk) and a pollen-bearing anther; individual unit of the third whorl of the flower (the androecium); the male sporophyll of flowers

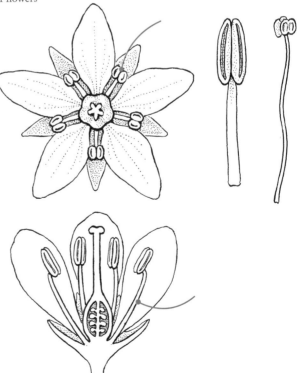

S

staminate

having male reproductive structures
(stamens) and lacking female structures
(pistils)

staminode

sterile stamen, often reduced in size (top) or
modified to attract pollinators (below)

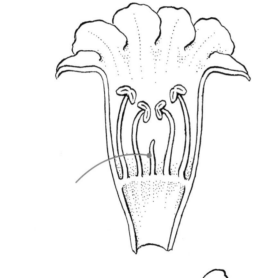

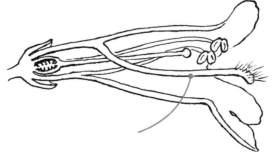

standard

1. one of the three inner tepals (all sepals)
in the flowers of irises (*Iris*); see also fall; 2.
flower petal typical of papilionoid legumes in
the bean family (Fabaceae), usually the upper
and largest petal, e.g., sweet peas (*Lathyrus*),
lupines (*Lupinus*)

SYNONYM 2. banner, vexillum

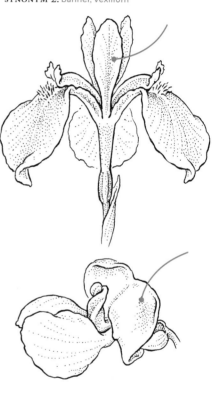

stellate

shaped like a star, most commonly used to
describe multiple hairs growing from the
same spot, like those found on hibiscus
(*Hibiscus*)

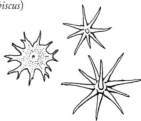

stem

the part of the plant from which leaves and buds emerge at nodes, usually above ground but sometimes below ground

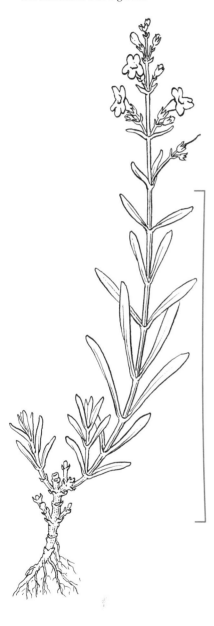

sterile

1. not currently in a reproductive state, i.e., not flowering or fruiting; 2. not capable of sexual reproduction, e.g., the vegetative fronds of sensitive fern (*Onoclea sensibilis*), or the showy florets of lacecap hydrangeas (*Hydrangea*)

SYNONYM 2. infertile

sticktight

plant or plant part that sticks to clothing, hair, or fur

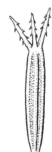

S

stigma

top section of the pistil that is receptive to
pollen; illustrated in detail below (clockwise
from upper left) are bifurcate, discoid,
plumose, and lobed stigmas

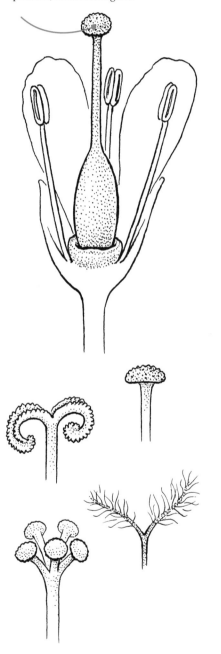

stilt root

adventitious root emerging from the lower
part of a trunk and acting as structural
support for a tree

SYNONYM anchor root, brace root, prop root

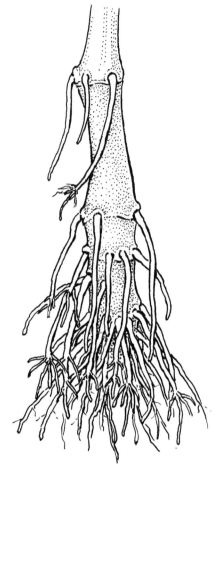

stipe

1. in ferns, the stalk of the frond, equivalent to the petiole of seed plants; 2. in orchids, the stalk-like connection between the viscidium and the pollinia

stipel

small leafy or spiny structure associated with the leaflet base or petiolule

stipellate

having stipels

stipular

pertaining to stipules, as with scars left on the stem when the stipules fall off, such as in the magnolia family (Magnoliaceae)

stipulate

having stipules

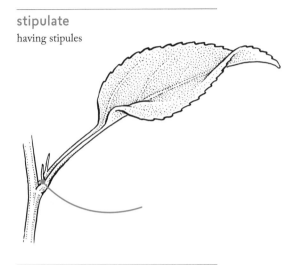

stipule

leafy or spiny structure associated with the petiole and/or node of some plants

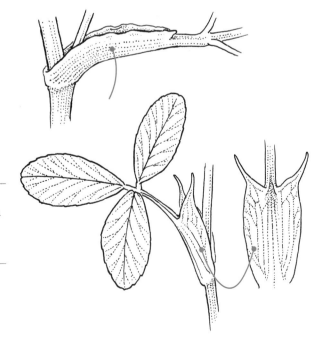

S

stolon

horizontal, above-ground, creeping stem with roots and shoots forming at nodes and the tip

SYNONYM runner

stone fruit

fleshy drupe with a hard endocarp

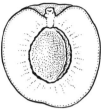

strap

long, narrow, fused corolla (ligule) of ray flowers in the sunflower family (Asteraceae)

striate

striped with lines, ridges, or grooves

strobilus

(plural strobili) cone-shaped, cylindrical, or spherical structure with a central axis on which scales bearing spores, pollen, ovules, or seeds (i.e., sporophylls) are attached; the reproductive structure of gymnosperms, some lycophytes, and a few angiosperms

stoloniferous

having stolons

stone

hard endocarp of fleshy drupes such as peaches and cherries (*Prunus*)

style
area of the pistil between the stigma and the ovary

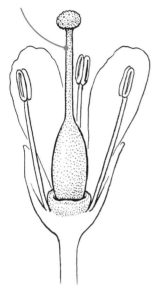

style arm, style branch
branch of a style, each usually with its own stigma

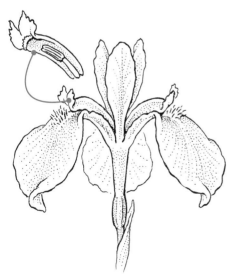

stylopodium
swollen disk-shaped area at the base of the style in the carrot family (Apiaceae)

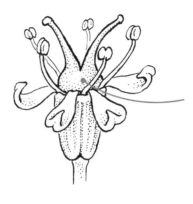

sub-
prefix meaning 1. nearly so, almost, not quite; 2. below, under, beneath

subfamily
taxonomic rank above genus and below family; plant subfamily names end in "-oideae"

submersed, submerged
growing fully below the water's surface
ANTONYM emersed

subshrub
small woody plant with multiple main stems

subspecies
taxonomic rank below species; individuals or populations usually have natural ranges and forms that differ from what is typical for the species but that do not warrant recognition as a distinct species; see also variety

S

subtend

occuring under another structure or organ; e.g., the epicalyx on a *Hibiscus* flower

subterranean

below ground

subulate

narrowly trowel-shaped or broadly needle-shaped

succulent

1. fleshy, water-storing; 2. plant that has fleshy, water-storing leaves and/or stems, e.g., jade plant (*Crassula ovata*), cacti (Cactaceae), spurges (Euphorbiaceae)

sucker

shoot growing from the base of a plant, usually applied to those emerging from below ground

suffrutex

subshrub, particularly one that is woody at the base and herbaceous above

summer annual

plant that grows from seed, flowers, produces seeds, and dies in the period from spring to early fall; see also winter annual

summer-bearing

a fruiting shrub with floricanes that produce fruit midway through their second year of growth; e.g., some raspberries and blackberries (*Rubus*); see also fall-bearing

super-

prefix meaning above something else, or exceeding normal

superior ovary

gynoecium that is attached above the points of attachment of the outer three floral whorls (calyx, corolla, androecium)

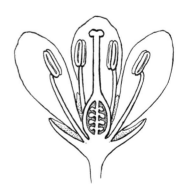

surculose
having or producing basal shoots or suckers

suture
line at which a fruit or anther splits open to dehisce

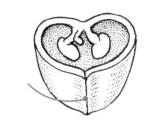

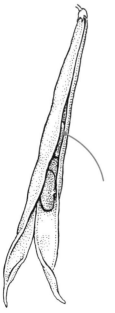

syconium
(plural syconia) enclosed inflorescence and multiple infructescence of figs (*Ficus*), consisting of an expanded receptacle folded in on itself and forming a chamber open only via an apical pore, flowers and resulting fused fruits are inside

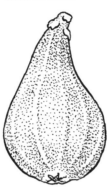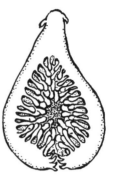

sym-
prefix meaning fused

symbiosis
a relationship in which two organisms live fused together or in very close proximity, usually beneficial to both (mutualistic)

sympatric
occurring in the same area, as with two species whose distributions overlap
ANTONYM allopatric

sympetalous
having a corolla that is at least partially fused
SYNONYM gamopetalous

S

sympodial

having a main axis made up of a series of terminating branches, each followed by an axillary branch that continues the outward growth of the axis; usually used to describe the growth of inflorescences but also to describe vegetative growth, such as in some orchids (Orchidaceae); see also monopodial

syn-

prefix meaning fused

synandrous

having anthers that are fused together, as in flowers of the sunflower family (Asteraceae)

syncarp

fruit derived from an entire inflorescence, may be fleshy or dry; e.g., sweetgum (*Liquidambar styraciflua*)

SYNONYM multiple infructescence

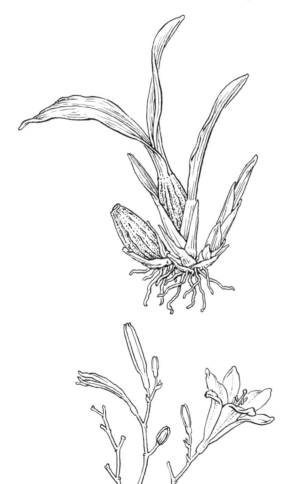

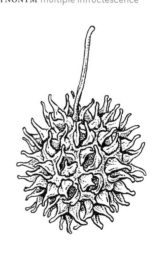

syncarpous

gynoecium consisting of two to many fused
carpels (a compound pistil)

ANTONYM apocarpous

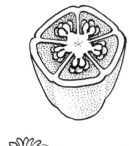

synsepalous

having a calyx that is at least partially fused

SYNONYM gamosepalous

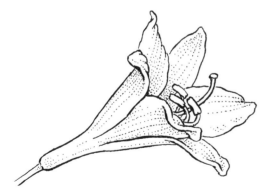

T

taproot

root system in which there is a primary root of much greater diameter than the lateral roots

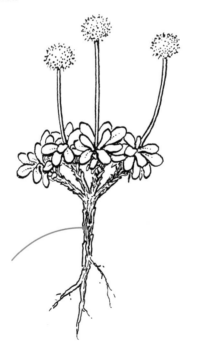

tassel

the apical, male inflorescence in corn (*Zea mays*)

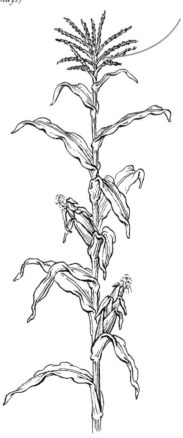

taxon

(plural taxa) a member of any taxonomic rank, such as a subspecies, species, genus, family, or order; useful in referring generically to a number of entities within a certain rank, e.g., the number of taxa in a genus would include species as well as subspecies and varieties

tendril

twining modified whole or partial stem, leaf, or leaflet that helps a plant attach to neighboring plants or other supports as it climbs

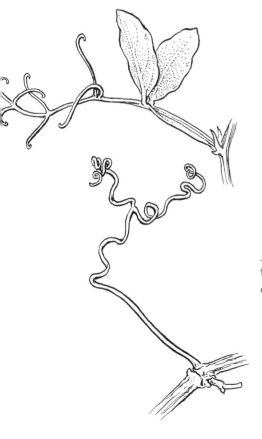

tepal

individual component of a perianth when the sepals and petals are petaloid; e.g., daffodils (*Narcissus*), daylilies (*Hemerocallis*)

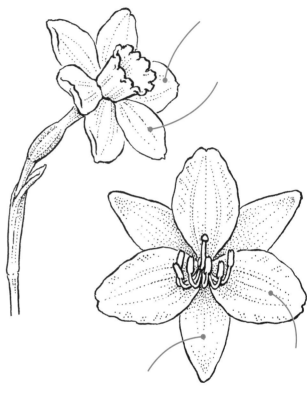

terete

circular in cross section

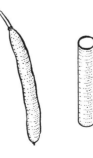

T

terminal

at the apex or tip, as of leaflets or inflorescence

terminal bud

bud located at the tip of a stem, responsible for the elongating growth of the stem in most woody plants

ternate

divided in three; e.g., trifoliolate leaf

terrestrial

growing in the ground; living in and dependent upon land, not a body of water

testa

layer of tissue covering a seed, derived from
the integuments surrounding the ovule

SYNONYM seed coat

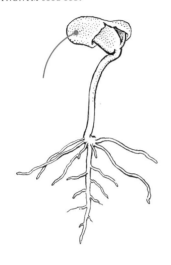

tetra-

prefix meaning four

tetradynamous

having six stamens, four of which are long
and two short, characteristic of the mustard
family (Brassicaceae)

tetragonal, tetrangular

having four angles, like the young stems
of most members of the mint family
(Lamiaceae)

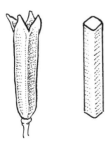

tetramerous

having flower parts in multiples of four

tetraploid

having four sets of chromosomes (4n); see
also diploid, haploid, polyploid

thallus

(plural thalli) main body of a plant that is not
differentiated into stems, roots, and leaves

theca

(plural thecae) one of two chambers inside of
each anther, usually bears pollen

SYNONYM anther sac

thigmotropism

growth or change in orientation in response
to touch

T

thorn

sharp, pointed modified stem

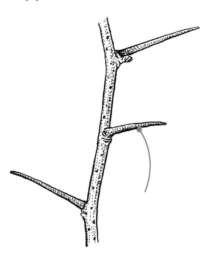

throat

in a fused corolla, the portion of the inside of the corolla that is seen when looking down into the flower

thyrse

branched inflorescence with primary branches borne along an elongate axis (racemose) and secondary branches cymose

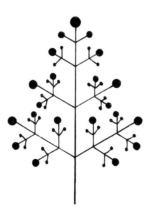

tiller

vertical shoot growing from the base or roots of a tree or shrub (i.e., a sucker)

tillering

propagation technique that harvests plant's basal vertical shoots (tillers) to cultivate as separate plants; the new plants are clones of the parent plant

tissue

cluster of like cells that together have a function

tissue culture

propagation technique that uses a small piece cut from the parent plant to grow a new adult plant on growth medium in a sterile environment; the new plant is a clone of the parent plant

tomentose

covered in short woolly hairs

tooth

serration or dentation along a margin

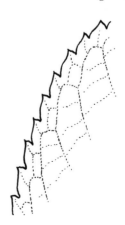

top
in horticulture, to prune away the upper
portion of a plant

torus
in flowers, tissue to which all the floral
whorls are attached
SYNONYM receptacle

trailing
having a horizontal stem creeping along the
ground but not rooted to it

translator, translator arm
narrow connector between the two pollen
clusters (pollinia) from different anthers in
milkweeds (*Asclepias*)

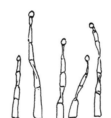

transplant
to move from one location to another

transverse
perpendicular to the main axis
SYNONYM latitudinal

tree
woody plant, generally with one trunk and
taller than mature shrubs

treelet
small tree

tri-
prefix meaning three

tribe
taxonomic rank above genus and below
family; plant tribe names end in "-eae"

tricarpelate
having three carpels

trichome
hair or similar outgrowth of the epidermis
consisting of one or more elongated cells

trifid
deeply divided into three lobes

trifoliate

having three leaves, often also used to refer to leaves with three leaflets (trifoliolate)

trifoliolate

having three leaflets

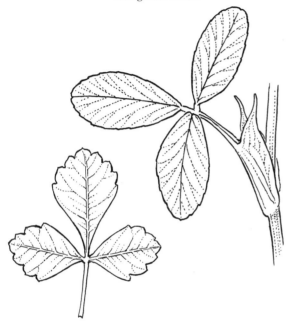

trigonous

triangular in cross section, as with a stem with three edges

trilobate

having three lobes

trilocular

having three locules (chambers) or cavities

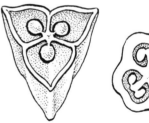

trimerous

having flower parts in multiples of three

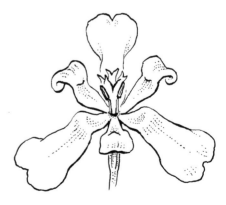

tripartite

divided into three parts

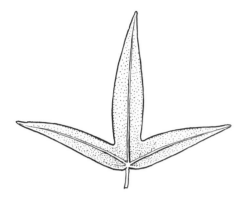

tropism

growth or change in orientation in response to a resource or stimulus

truncate

straight, as if the apex or base has been cut off with a straight edge (squared)

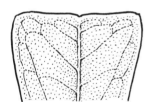

trunk

main stem or axis of a tree, between the roots and where branches begin to form the crown

SYNONYM bole

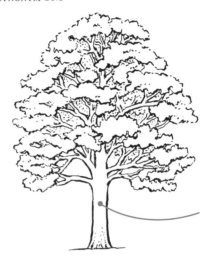

truss

in horticulture, a cluster of flowers, an inflorescence

tuber

a thickened rhizome (underground stem), with nodes and internodes, that stores energy, usually in the form of starch; term is often used to refer to the individual components of tuberous roots, which are true roots

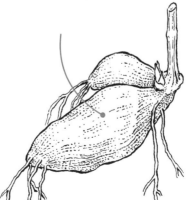

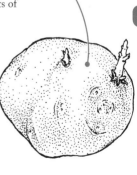

T

tubercle

small protuberance that resembles a tuber;
e.g., the achene of sedges (Cyperaceae)

tuberous roots

swollen roots in which energy is stored,
usually in the form of starch

tubular

cylindrical, resembling a tube

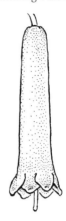

tufted

occurring in dense little clusters

tunicate

having multiple, concentric layers, as with the
leaves of onion (*Allium*) bulbs

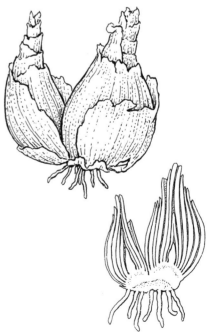

turbinate

shaped like a spinning top, as with the flower buds of dogwood (*Cornus florida*)

turgid

swollen, distended, often due to water absorption

turion

a bud in aquatic plants that acts as a vegetative propagule; turions sink to the bottom of water, surviving harsh winters and drought, and rise again when conditions are favorable for growth

tussock

clump of grass or grass-like plants that is larger than other similar plants around them, as with a tuft of grass that has grown taller than the rest of a lawn

twig

small, somewhat delicate woody stem

twining

coiling or turning around something else for support

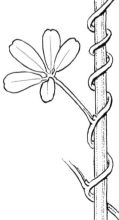

two-ranked

occurring in a single plane along a central axis, as with leaves on a stem, making the entire structure appear flat when viewed down the axis from the tip to the bottom or vice versa

SYNONYM distichous

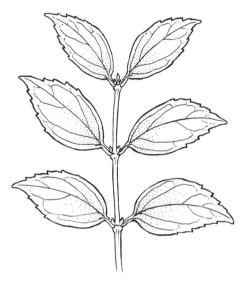

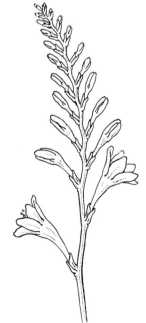

T

U

ubiquitous

distributed worldwide, or nearly so

SYNONYM cosmopolitan

umbel

inflorescence with flowers or branches borne from a common point, presenting flowers on a rounded or flat plane at the top, may be simple (unbranched) or compound (branched); illustrated here are shootingstar (*Dodecatheon*) and onion (*Allium*), both simple, and Queen Anne's lace (*Daucus carota*), compound

umbo

projection on the outside surface of female cone scales (megasporophylls)

unarmed

without prickles, spines, or thorns

uncinate

hooked at the tip, as with some tendrils and leaves

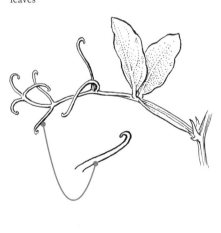

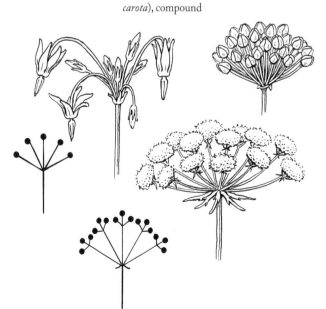

undulate

margin or surface shallowly wavy, more often applied to surfaces

SYNONYM repand

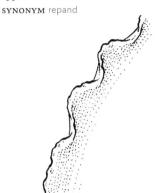

unguiculate

clawed

uni-

prefix meaning one

unicarpellate, unicarpellous

having only one carpel

unifoliate

having only one leaf

unifoliolate

having compound leaves that are reduced to one leaflet and thus appear to be simple

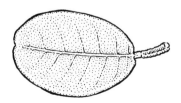

unilocular

having only one locule (chamber) or cavity

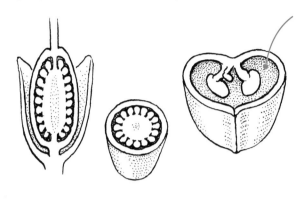

uniseriate

having parts arranged in one row

unisexual

having only male or only female functional reproductive parts

urceolate

shaped like an urn

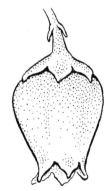

U

V

valvate

1. having petals or sepals that line up margin to margin in bud, not overlapping; 2. opening by valves, as with some capsules and anthers

valve

one of the separating segments in a dehiscing fruit

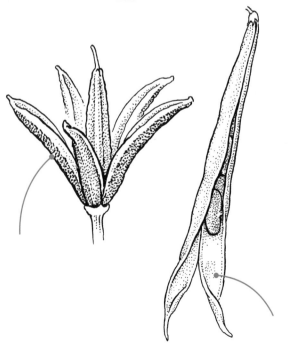

variegated

having more than one color in organs or tissues that are normally a solid color (usually green), most often applied to leaves or whole plants such as coleus (*Solenostemon*)

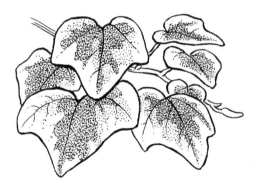

variety

taxonomic rank below species; individuals or populations usually have a certain stable but minor characteristic, such as flower color, that differs slightly from what is typical for the species; see also subspecies

vascular bundle
tight columns of conductive tissue, traces of these bundles are visible in leaf scars left when leaves fall from woody plants

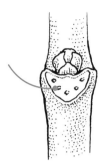

vascular tissue
cells that together transport water and nutrients throughout the plant bodies of seed plants, ferns, and lycophytes

vegetative
non-reproductive parts of plants; e.g., stems, leaves, roots

vein
vascular tissue in leaves or leaf-like structures such as bracts, petals, sepals, and stipules, may be branching or not
SYNONYM nerve

velamin
the spongy, water-absorbing outer epidermal layer of epiphytic orchid roots

velutinous
covered in short velvety hairs

venation
the arrangement of vascular tissue in leaves or leaf-like structures such as petals and sepals

ventral
pertaining to the front, surface facing toward the axis
ANTONYM dorsal

ventricose
inflated on only one side, usually in the middle

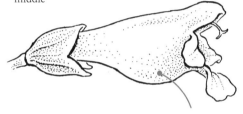

vernal
of the spring, as with plants that bloom in the spring

vernation
arrangement of leaves in bud; see also aestivation
SYNONYM ptyxis

versatile
attached at the middle, as with filaments attached to the middle of anthers; see also basifixed, dorsifixed
SYNONYM medifixed

V

verticil

one of several layers around a central axis, as with parts of a flower; e.g., corolla, calyx, androecium, gynoecium

SYNONYM whorl

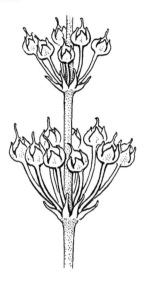

verticillaster

inflorescence of opposite cymes that occur in a series of pairs along the terminal portion of stems, creating a false whorl, as in the mint family (Lamiaceae)

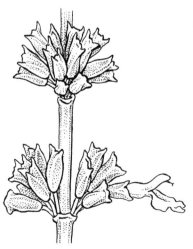

verticillate

1. occurring more than two per node, as with multiple leaves circling a stem; 2. arranged in layers around a central axis

SYNONYM whorled

vestigial

underdeveloped, reduced in size and not functional; e.g., non-functional, reduced stamen (staminode) in a flower

SYNONYM obsolete, rudimentary

vestiture

the collective covering of and projections from a plant's epidermis

vexillum

flower petal typical of papilionoid legumes in the bean family (Fabaceae), usually the upper and largest petal; e.g., sweet peas (*Lathyrus*), lupines (*Lupinus*)

SYNONYM banner, standard

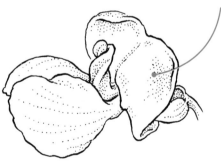

viable

capable of life or reproduction, as with a seed that germinates and grows into a seedling

villous

covered in soft, long hairs that remain
untangled

vine

herbaceous climbing plant

viscidium

sticky structure attached to an orchid
pollinium via stipes, the viscidium facilitates
pollen transfer by sticking to pollinators

vitreous

glass-like in appearance, transparent

viviparous

offspring developing while still attached to
the parent plant, as with seeds germinating in
fruits still attached to the parent plant (e.g.,
mangroves, *Avicennia*, *Rhizophora*), or buds
that forms plantlets while still attached to the
parent plant (often on leaves, e.g., waterlilies,
Nymphaea)

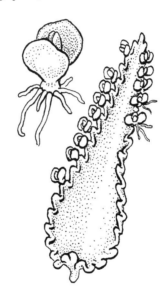

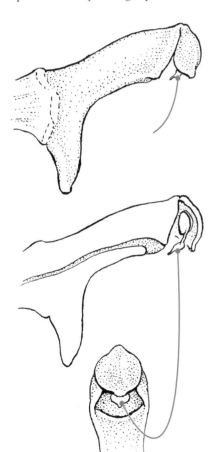

W

weed

a plant that grows where it is not wanted and is difficult to eradicate, usually applied to plants in cultivated or otherwise disturbed places

weeping

having branches that hang downward, as with some willows (*Salix*)

whorl

one of several layers around a central axis, as with parts of a flower; e.g., corolla, calyx, androecium, gynoecium
SYNONYM verticil

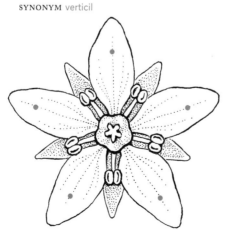

whorled

1. occurring more than two per node, as with multiple leaves circling a stem; 2. arranged in layers around a central axis
SYNONYM verticillate

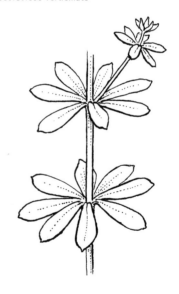

wing

1. flat flap or extension of tissue emerging from the margin of a structure, such as a leaf rachis or a fruit; 2. lateral petal of a pea flower (Fabaceae subfamily Papilionoideae)

winter bud

dormant plant shoot covered in scales that protect it from environmental conditions such as frost

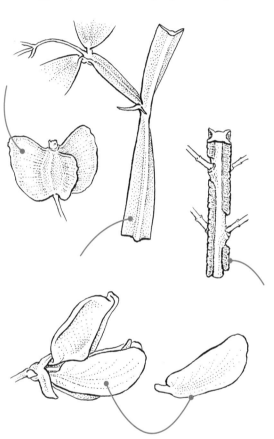

winter annual

plant that grows from seed, flowers, produces seeds, and dies in the period from early fall to late spring; see also summer annual

X

in nomenclature, indicates hybrid origin, whether written before the genus, indicating an intergeneric cross (e.g., ×*Heucherella*), or between the genus and the specific epithet, indicating an interspecific cross (e.g., *Epimedium* ×*versicolor*)

xanthophyll

yellow pigments primarily in plant leaves, oil-soluble

xeric

dry, pertains to geographic areas such as deserts

xero-

prefix meaning dry

xerophyte

plant adapted to growing with very low water availability; see also hydrophyte, mesophyte

X.S.

cross section

ANTONYM l.s., longitudinal section

Z

zoophilous

pollinated by animals, especially those other than insects

zygomorphic

having a single plane of symmetry such that only one line drawn through the middle produces two mirror-image halves

SYNONYM bilaterally symmetrical; irregular

ANTONYM actinomorphic, radially symmetrical, regular

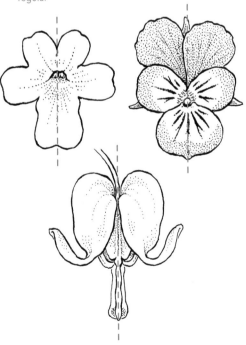

Z

Recommended Reading

Bebbington, Anne L. D. 2015. *Understanding the Flowering Plants: A Practical Guide for Botanical Illustrators.* The Crowood Press, Marlborough, U.K.

Beentje, Henk. 2010. *The Kew Plant Glossary: An Illustrated Dictionary of Plant Terms.* Royal Botanic Gardens, Kew, London, U.K.

Bell, Adrian D. 2008. *Plant Form: An Illustrated Guide to Flowering Plant Morphology.* Timber Press, Portland, Ore.

Castner, James L. 2004. *Photographic Atlas of Botany and Guide to Plant Identification.* Feline Press, Gainesville, Fla.

Ellis, Beth, Douglas C. Daly, Leo J. Hickey, John D. Mitchell, Kirk R. Johnson, Peter Wilf, and Scott L. Wing. 2009. *Manual of Leaf Architecture.* Comstock Publishing Associates, an imprint of Cornell University Press, Ithaca, N.Y.

Gough, Robert. E. 1993. *Glossary of Vital Terms for the Home Gardener.* Food Products Press, an imprint of The Haworth Press, Inc., Binghamton, N.Y.

Harris, James G., and Melinda Woolf Harris. 2001. *Plant Identification Terminology: An Illustrated Glossary*, 2nd ed. Spring Lake Publishing, Spring Lake, Utah.

Hickey, Michael, and Clive King. 2001. *The Cambridge Illustrated Glossary of Botanical Terms.* Cambridge University Press, Cambridge, U.K.

Horticultural Research Institute. 1971. *A Technical Glossary of Horticultural and Landscape Terminology.* Pennsylvania State University, Department of Landscape Architecture, University Park, Pa.

Mabberley, David J. 2008. *Mabberley's Plant-book: A Portable Dictionary of Plants, Their Classifications, and Uses*, 3rd ed. Cambridge University Press, Cambridge, U.K.

Swartz, Delbert. 1971. *Collegiate Dictionary of Botany.* The Ronald Press Company, New York, N.Y.

Zomlefer, Wendy B. 1994. *Guide to Flowering Plant Families.* The University of North Carolina Press, Chapel Hill.

About the Authors

Susan K. Pell is the Science and Public Programs Manager at the United States Botanic Garden, where she gets to show people the awesomeness of plants every day. She was formerly Director of Science at the Brooklyn Botanic Garden, where she studied the evolutionary relationships of the cashew family. She holds a Ph.D. in plant biology and teaches courses in genetics, angiosperm morphology, and systematics. Susan lives in Washington, D.C., with her wife and daughter.

ALLISON MILLER

Bobbi Angell creates richly detailed pen and ink drawings for botanists at the New York Botanical Garden and other institutions, and for many years illustrated the *New York Times* Garden Q&A column. A gardener and printmaker as well as an illustrator, she lives in southern Vermont.

MALLORY LAKE